Sustainability Science: Field Methods and Exercises

Miguel Esteban · Tomohiro Akiyama
Chiahsin Chen · Izumi Ikeda · Takashi Mino
Editors

Sustainability Science: Field Methods and Exercises

Springer

Editors
Miguel Esteban
Graduate Program in Sustainability Science-Global Leadership Initiative, Graduate School of Frontier Sciences
The University of Tokyo
Kashiwa
Japan

Tomohiro Akiyama
Graduate Program in Sustainability Science-Global Leadership Initiative, Graduate School of Frontier Sciences
The University of Tokyo
Kashiwa
Japan

Chiahsin Chen
Graduate Program in Sustainability Science-Global Leadership Initiative, Graduate School of Frontier Sciences
The University of Tokyo
Kashiwa
Japan

Izumi Ikeda
Graduate Program in Sustainability Science-Global Leadership Initiative, Graduate School of Frontier Sciences
The University of Tokyo
Kashiwa
Japan

Takashi Mino
Graduate Program in Sustainability Science-Global Leadership Initiative, Graduate School of Frontier Sciences
The University of Tokyo
Kashiwa
Japan

ISBN 978-3-319-32929-1 ISBN 978-3-319-32930-7 (eBook)
DOI 10.1007/978-3-319-32930-7

Library of Congress Control Number: 2016938393

Printed on acid-free paper

This Springer imprint is published by Springer Nature
The registered company is Springer International Publishing AG Switzerland

Preface

The last couple of decades have seen a fundamental shift in how society perceives it should prepare for the future, giving rise to the discipline of sustainability science. Since the issues we are facing today are highly complex, such as climate change and difficulty in meeting the rising energy demands while not harming the planet, it is important that they are dealt with in an interdependent and holistic manner. Inevitably, this requires academia to undergo a transformation from following a sectionalized approach to the one where different disciplines and fields collaborate together, essentially a transdisciplinary approach.

Since establishing the Integrated Research System for Sustainability Science (IR3S) in 2005 and the Graduate Program in Sustainability Science (GPSS) in 2007, the University of Tokyo has become a widely recognized leader not only in advancing sustainability research, but also in attempting to apply in practice the findings from such research. Building on the foundations and progress forged by the IR3S and GPSS, "the Graduate Program in Sustainability Science-Global Leadership Initiative (GPSS-GLI)" was established in 2011 to advance the field of sustainability science by aiming to train the next generation of "global leaders". Such leaders should be characterized as individuals that are not only highly specialized in their own fields, but also have extensive knowledge of a variety of other disciplines and are guided by ethically sound principles. Essentially, GPSS-GLI is one of the nine competitive degree programs within the University of Tokyo that is being supported by "Program for Leading Graduate Schools" initiative funded by Japan's Ministry of Education, Culture, Sports, Science and Technology (MEXT). As a collaborative effort between the Graduate School of Frontier Sciences and the United Nations University (UNU), GPSS-GLI combines the educational resources and international research networks of these leading institutions and thereby provides participants with the training and opportunities necessary to become global leaders.

One of the key elements in the GPSS-GLI program to train "global leaders" is to provide students with ample opportunity to experience the reality in the field, framed around Global Field Exercise (GFE) and Exercises on Resilience (ER). The exercises not only help students broaden their horizons and attempt to holistically understand problems and develop solutions, but also serve to develop general methodologies that students and sustainability science practitioners can use in the field. This book attempts to summarize some of the experiences in running these GPSS-GLI courses and showcase other field works that GPSS-GLI students have undertaken as part of their formation as sustainability science leaders.

We hope that the book will serve as a good source of background information for those who wish to conduct field exercises in sustainability science, by illustrating the type of research that is possible, and inspire others to continue to develop conceptual and practical ways of conducting such work.

Miguel Esteban
Tomohiro Akiyama
Chiahsin Chen
Izumi Ikeda
Takashi Mino

Contents

Part I Theories, Concepts and Methodologies in Sustainability Science

Philosophy of Field Methods in the GPSS-GLI Program: Dealing with Complexity to Achieve Resilience and Sustainable Societies 3
Takashi Mino, Miguel Esteban, Vivek Anand Asokan, Niranji Satanarachchi, Tomohiro Akiyama, Izumi Ikeda and Chiahsin Chen

Part II Global Field Exercises

Designing Field Exercises with the Integral Approach for Sustainability Science: A Case Study of the Heihe River Basin, China 23
Ricardo O. San Carlos, Heng Yi Teah, Tomohiro Akiyama and Jia Li

Field Survey Key Informant Interviews in Sustainability Science: Costa Rica's PES Policy of Changing Focus from Quantity to Quality 41
Doreen Allasiw, Yuki Yoshida, Giles Bruno Sioen, Rene Castro, Ying Palopakon, Toshinori Tanaka, Toru Terada, Akiko Iida and Makoto Yokohari

Part III Exercises on Resilience

Assessment of Fieldwork Methodologies for Educational Purposes in Sustainability Science: Exercise on Resilience, Tohoku Unit 2015 67
Ricardo O. San Carlos, Olga Tyunina, Yuki Yoshida, Aimee Mori, Giles Bruno Sioen and Jiaqi Yang

Drawing Lessons from the Minamata Incident for the General Public: Exercise on Resilience, Minamata Unit AY2014 93
Eri Amasawa, Heng Yi Teah, Joanne Yu Ting Khew, Izumi Ikeda and Motoharu Onuki

Part IV Sustainability Science Field Research

Sustainability Science as the Next Step in Urban Planning and Design 117
Giles Bruno Sioen, Toru Terada and Makoto Yokohari

A Methodology to Evaluate Sustainability in the Face of Complex Dynamics: Implications for Field Studies in Sustainability Science 137
Niranji Satanarachchi and Takashi Mino

Sustainability Field Exercises in Rural Areas: Applying the Community Marginalization Framework to Examine Qualitative Changes in Rural Communities 153
Shogo Kudo

Participatory Mapping and Problem Ranking Methodology in the Research of Sustainable Communities—Workshop with Indigenous People Under Community-Based Forest Management Program in the Philippines 177
Marcin Pawel Jarzebski

Rapid Sustainability Appraisal of Collapsed Jatropha Projects in Ghana Using Local Community Perceptions: Methodological Implications for Sustainability Science 199
Abubakari Ahmed and Alexandros Gasparatos

Index 229

Part I
Theories, Concepts and Methodologies in Sustainability Science

Philosophy of Field Methods in the GPSS-GLI Program: Dealing with Complexity to Achieve Resilience and Sustainable Societies

Takashi Mino, Miguel Esteban, Vivek Anand Asokan, Niranji Satanarachchi, Tomohiro Akiyama, Izumi Ikeda and Chiahsin Chen

Abstract The world is facing a multitude of pressing problems, including environmental degradation, natural disasters, and social inequity, to name but a few. These challenges are also complex and uncertain in nature, though it is crucial for humanity to attempt to solve them in order to achieve sustainable societies. The Graduate Program in Sustainability Science-Global Leadership Initiative (GPSS-GLI) of the University of Tokyo is an academic program which looks forward to facing these challenges. The program has a strong focus on field exercises, which attempt to introduce students to the real situations being experienced by people. Students are encouraged to deal with complexity by engaging the issue from a holistic ("top-down") and transboundary ("bottom-up") perspective. Having a holistic view and transboundary perspective may provide a basis to deal with the complexities and uncertainties present in sustainability issues, where it is difficult to provide solutions by thinking only of fixed end-targets. Through such efforts it is hoped that students can understand and propose solutions on how to achieve more sustainable and resilience societies. The present chapter will serve as an introduction to the rest of the chapters in this book, briefly outlining the general philosophy of the GPSS-GLI regarding Global Field Exercises (GFEs) and Exercises in Resilience (ERs).

Keywords Resilience · Sustainable societies · Field methods · Sustainability science · Philosophy · GPSS-GLI

1 Introduction

Humanity faces an array of serious and complex issues in the 21st century, including climate change, resource depletion, biodiversity loss, large-scale disasters,

---

T. Mino · M. Esteban (✉) · V.A. Asokan · N. Satanarachchi · T. Akiyama · I. Ikeda · C. Chen
Graduate Program in Sustainability Science-Global Leadership Initiative,
Graduate School of Frontier Sciences, The University of Tokyo,
5-1-5 Kashiwanoha, Kashiwa, Chiba 277-8563, Japan
e-mail: esteban.fagan@gmail.com

M. Esteban et al. (eds.), *Sustainability Science: Field Methods and Exercises*,
DOI 10.1007/978-3-319-32930-7_1

energy crises, widespread poverty, financial insecurity, or rapid population changes. The future health of ecosystems and society—the very things that ensure the survival of humanity—depend on how we forge pathways of sustainable development, and require a new generation of leaders that pair extensive knowledge and intensive specialization with a critical perspective and strong ethics.

The University of Tokyo has taken up the challenge of improving global sustainability through the use of transboundary and holistic approaches, as shown in Table 1. With the establishment of the Integrated Research System for Sustainability Science (IR3S) and the Graduate Program in Sustainability Science (GPSS), the University of Tokyo became a widely recognized leader in sustainability research and applying research findings in practical ways through collaborative partnerships with external actors and institutions. In 2012, GPSS started a new education project called the Graduate Program in Sustainability Science–Global Leadership Initiative (GPSS-GLI) to develop individuals with extensive knowledge, intensive specialization, and ethically sound principles who can form the next generation of global leaders.

As a result, throughout its Masters and PhD programs, GPSS-GLI aims for students be able to "develop the skills necessary for global leadership, to acquire a broad perspective and problem solving capabilities, and to learn to apply the concept of "resilience" both theoretically and practically" (GPSS-GLI 2015). To achieve these objectives the program emphasizes the importance of achieving a holistic understanding, looking to build a resilient society that can absorb shocks and approaches issues in a transboundary manner. With the help of various collaborative partners, GPSS-GLI combines the educational resources and international

Table 1 Milestones in research and education in sustainability at the University of Tokyo

1996	Alliance for Global Sustainability (AGS) established (http://en.ags.dir.u-tokyo.ac.jp)
2000	Youth encounter on sustainability (YES) started
2004	Intensive program on sustainalility (IPoS) set up
2005	Intensive research system for sustainability science (IR3S) established as first research institute at University of Tokyo (http://en.ags.ir3s.u-tokyo.ac.jp)
	International alliance for research universities (IARU) created (http://www.iaruni.org)
2006	Transdisciplinary initiative for global sustainability (TIGS) launched (http://en.ags.dir.u-tokyo.ac.jp/)
2007	Grduate program in sustainability science (GPSS) established on Graduate School of Frontier Sciences (GSFS) and jointly operated by six departments within Division of Environmental Studies: Environment Systems, Human and Engineered Environmental Studies, International Studies, Ocean Technology, Policy and Environment Natural Environmental Studies, and Socio-Cultural Environmental Studies
	International academic journal *Sustainability Science* (Springer) created
2008	Asian program for incubation of Environmental leaders (APIEL) started (http://www.envleader.u-tokyo.ac.jp/index_e.html)
2009	First international conference on sustainability science (ICSS) hosted by IR3S. Since then, ICSS and its Asian version, ICSS-Asia, have been hosted annually by IR3S
2010	Sustainability science consortium (SSC) created (http://ssc-g.net; in Japanese only)
2012	International society for sustainability science (ISSS) established (http://www.sussci.org)
	GPSS-GLI launched

research networks to provide students with the training and opportunities necessary to become global leaders. Through foundation courses and intensively specialized studies, as well as international and hands-on experience, students develop the skills necessary for global leadership, acquire a broad perspective and problem-solving capabilities, and learn to apply the concepts of sustainability and resilience both in theory and practice. Central to this curriculum is the idea that the best way to learn about the complexity inherent in modern societies is for students to experience it first-hand, by observing the situation in the field in order to attempt to obtain a complete view of the issues involved.

To this aim, a number of Field Exercises courses have been run in GPSS-GLI within the past few years. These exercises are continuously changing, as staff members strive to constantly improve them, and are held in a variety of countries and environments, in order to provide ample opportunity for students to experience a wide variety of situations. Students in the GPSS-GLI are very diverse, coming from all continents and disciplines, and thus holding exercises in a wide variety of environments allows them to be exposed to issues that they had not previously encountered in their home country or original discipline.

Nevertheless, despite the disparity in the types of field exercises held, there are a number of core principles and guiding philosophy that holds the GPSS-GLI program together. The present chapter thus serves as an introduction to the other chapters within this book, attempting to explore what are the principles and guiding philosophy behind field exercises that are run by GPSS-GLI. Though this exercise the authors hope that sustainability scientist practitioners and educators in other universities will be able to gain some insight into the ideas typically being discussed and implemented in GPSS-GLI, and that this will help to further the field of sustainability science in general.

2 Field Exercise Courses in the GPSS-GLI

The research approach of sustainability science is often recognized as having a problem-oriented perspective as its starting point (Lang et al. 2012; Swart et al. 2004; Clark and Dickson 2003; Kates et al. 2001). In addition, the importance of developing a contextual understanding and how this capacity can be gained by sustainability scientists have also been recognized (Mino and Hanaki 2013). Possessing a problem-oriented perspective and focusing on contextual importance means that sustainability research relies heavily on field studies, allowing its practitioners to obtain a first-hand experience of the situation.

To provide a platform to help students gain such skills, GPSS-GLI runs two different types of exercise courses, known as Global Field Exercises (GFE) and Exercises on Resilience (ER). Each of these is typically organized by 2 or 3 academic members of staff, often assisted by one PhD student who can obtain credits through a Global Leadership Exercise (GLE) course. The overall purpose of this GLE is to develop PhD students' leadership skills, and thus they are given partial responsibility in the planning, implementing, and evaluating of the GFE. Field-oriented exercises can be considered to effectively develop leadership, through activities such as student-based group work, which include reviewing literature, planning on-site research activities, implementing field surveys and presenting the results of the study (Akiyama et al. 2012).

Each GFE or ER consists roughly of four parts: preliminary learning (2–6 months), on-site learning in the field (typically 10 days–2 weeks), after-the-fact learning, and a joint task (report preparation and a presentation).

In the present book 2 GFE and 2 ER courses will be introduced. However, in addition to these field exercises, essentially course modules, many students in GPSS-GLI engage in field research for their Masters or PhD thesis work. It should be noted that there are many instances in which the field exercise courses at GPSS-GLI greatly influence and shape the students' individual research, and thus the importance of these field exercises cannot be underestimated. To showcase some of these examples the later chapters of this book will introduce field work activities and conceptual thinking by a number of current and former students of the program, illustrating the contribution of educational activities in GPSS-GLI to the development of field methodologies in sustainability science.

2.1 Global Field Exercise (GFE)

GFEs take place several times each year, in cooperation with collaborating partners in Asia and Africa. Various "units" are created and students have to apply to join them, with a selection process determining which students are allocated to each GFE. GFEs are intended to broaden students' perspectives and cultivate an on-the-ground competency to identify issues through various activities, including preliminary surveys, site visits, experimental studies, discussions with various stakeholders (including local researchers and administrators), engagement in group work activities, and compilation and presentation of reports. Students are asked to adopt a solution-oriented approach that should holistically consider all the issues presented, in order not to provide generalistic solutions that could lead to future problems. The focus should always be the development of systems ("systems approach") that lead to more sustainable and resilient societies.

2.2 Resilience Exercises (ER)

The sustainability of our lifestyles is threatened by long-term environmental shifts, such as climate change, natural calamities, human-made disasters, and environmental destruction. Resilience, the ability to recover from such external disturbances, is a crucial factor in building a sustainable society. The Exercises on Resilience (ER) were designed as a specific part of the GPSS-GLI curriculum, focusing on giving students a fieldwork experience that would contribute to their understanding of the concept, applied to real issues in sustainability. ERs address the development of a resilient society through hands-on work examining reconstruction projects related to the 2011 Tohoku Earthquake tsunami as well as studies of past man-made disasters, such as that of the Minamata disease in Japan.

3 Complex Sustainability Issues

The GPSS-GLI tries to foster in students holistic and trans-boundary thinking, with the aim that they will later help to create more resilient societies. Clearly, this is a rather ambitious goal, and in the next section the authors will explain in more detail the philosophy of the program and how it is taught. Essentially, during field exercises (be it GFEs or ERs) students are taught to start to look at any situation by attempting to recognise complexity. In order to then analyse the various issues identified, it is necessary to adopt a holistic ("top-down") and transboundary ("bottom-up") approach, which can help to deal with issues of uncertainty and unpredictability. The outcome of such exercises can help students identify potential pathways to achieve more sustainable and resilient societies, as summarised in Fig. 1.

3.1 Complexity

One of the key characteristics of Field Exercises in GPSS-GLI is how they attempt to convey to students the inherent complexity that exists in any given situation that can be analysed from a sustainability point of view. Complexity, which in itself is a complex idea, lacks a single definition (Page 2010), and refers to a system with many parts, feedbacks, non-linear and linear relationships (Ladyman et al. 2013). The biological world and the socio-economic world are filled with smaller structures within a bigger scheme. These consist of many components which are interconnected via multiple pathways, giving rise to a complex system. (Peter and Swilling 2014; Levin 2006).

Essentially, in any given problem there are a variety of time scales (temporal diversity, such as multiple generations), spatial scales (such as global and local issues), governance scales (such as global, national or prefectural levels) and so on. Whether a problem is looked at from a national, societal, community or institutional points of view is likely to lead to different assessments, and these different scales make sustainability problems complex. For example, Chap. 4 highlights that there are a variety of time scales regarding reconstruction following the 2011 Tohoku Earthquake Tsunami, depending on whether one focuses on the short-term reconstruction process or the long-term viability of the community, which brings in issues of population aging and decline. The problems at the local and prefectural levels are

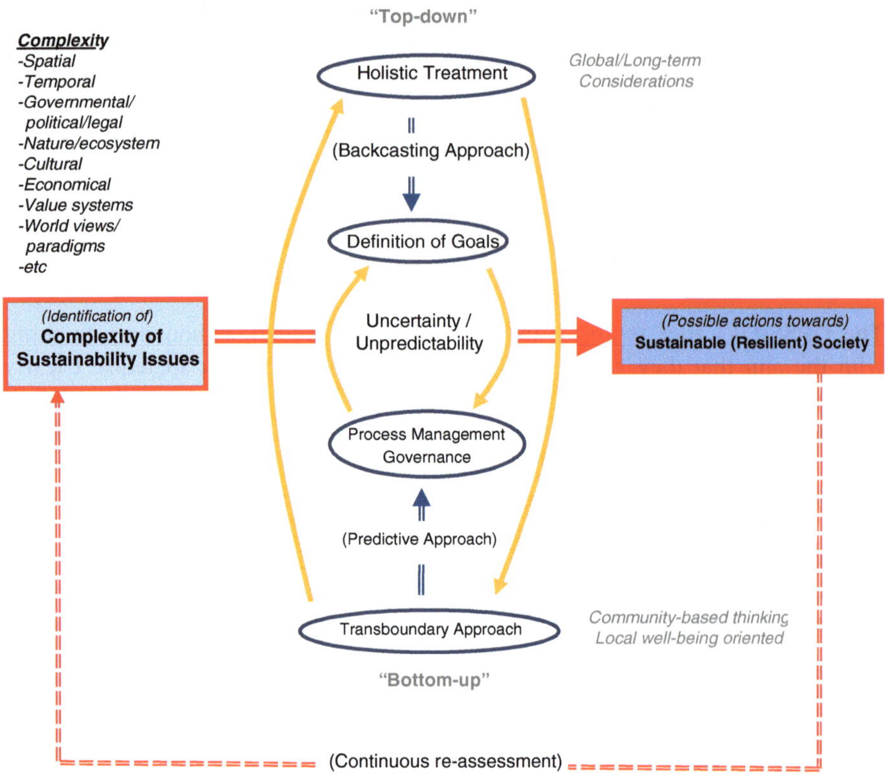

Fig. 1 Addressing complex sustainability issues: thought pathways behind GPSS-GLI's field exercises

also not the same, as different size communities are facing a variety of reconstruction issues, highlighting also how the concerns of the prefectural government might not be well aligned to those at the city or town level (Esteban et al. 2015).

There is also the diverse nature of stakeholders (with particular cultural and political points of view, and their own value chains), diverse ecosystems and natural conditions, each of which has to be taken into account. This can be exemplified by the contradicting views that can be expressed during a reconstruction process, were the interests of factory owners might be different to the wishes and aspirations of young people and NGOs (see Chap. 4 and Esteban et al. 2015). Also, reconstruction processes will depend on where they take place, and it is clear that the environmental and ecosystem constraints of rebuilding in the Sanriku coastline are different to those in coastal plains south of Sendai (both in the Tohoku region affected by the 2011 tsunami, see Mikami et al. 2012). In this sense, Clark identifies that sustainability should "understand the complex dynamics that arise from interactions between human and environmental systems" (Clark 2007), which is clearly dependent on the geography of a particular area. This means that when tackling sustainability issues it is important to reach a sound contextual understanding

(Agrawal 2008; Robinson 2011). It is important to also remember that each location does not constitute an isolated entity, but generate complexities that spans across different scales, and therefore it is essential that they are taken into account.

In addition, sustainability issues often regard diverse ecosystems and diverse natural conditions, and their influence on human well-being. Tackling sustainability issues in human-natural systems requires one to be aware of the complexities and dynamics of these systems (Liu et al. 2007; Ostrom 2007; Holling et al. 2002). Aside from acknowledging the complex nature of the systems and issues, we must also acknowledge that there are bottlenecks in recognizing and interpreting sustainability in the face of complexity, mainly due to limitations in the observation processes that one would adopt in the field (see Chap. 7 for a more detailed discussion). For instance, during a given field exercise a variety of issues can be identified, which are typically classified into the three dimensions of sustainability: social, environmental and economic (Lozano 2008), also known as the three pillars of sustainability. However such well-known classification alone would not be enough to comprehend the actual complexity of the issues that are encountered in the field. Over the years the three pillar view of sustainability has been expanded to include dimensions such as institutions, ethics, culture etc. (Hawkes 2001; Gibson et al. 2000). Others have also addressed the diversity found in the concept of sustainability (Neumayer 2003; Hopwood et al. 2005; Bell and Morse 2008; Espinosa et al. 2008), and alternative ways of structuring the concept have been proposed (Komiyama and Takeuchi 2006). Field research is one instance where the complexities of these conceptual interpretations and the complexities on the ground closely interact. Finally, in order to adequately recognize the complex nature of sustainability issues in the field, a sustainability scientist should also be aware of factors such as incomplete knowledge and multiple value systems.

3.2 Complex Issues Need Holistic Treatments ("Top-Down")

Linear thinking to come up with a solution to a given problem often leads to the creation of other future problems that will eventually have to be solved. A classic example from the field of coastal engineering involves the building of groynes to stop coastal erosion and restore beaches, which leads to exacerbated erosion further down the longshore drift direction (an example from Vietnam is discussed in Takagi et al. 2014). Such simplistic solutions, looking only at a problem in a narrow sense, can cause greater harm to the larger society in the long-term, highlighting the need to holistically look at any situation.

GPSS-GLI aims to foster in its students an in-depth and broad understanding of the complexity of human-natural systems as well as their dynamics. This is done by attempting to encourage the students to have a systemic, holistic view towards the issues which they research. Systems thinking and adopting a systemic view is well recognized as useful in sustainability research (Clayton and Radcliffe 1996; Bell and Morse 2008). However, traditional research on sustainability has also focused on solving a given problem from a particular perspective (Komiyama and Takeuchi 2006). Often, sustainability scientists and practitioners tend to focus on one or a few aspects of sustainability rather than explore their connectivity to other aspects in a holistic manner. This is contrary to the initial objectives of sustainability science, which aimed to bring together various perspectives, methodologies and approaches from both natural science and social science (Kates et al. 2000; Clark and Dickson 2003; Clark 2007). Recently Spangenberg (2011) re-emphasized how the science of sustainability should include not only natural science and economics, but also social sciences and humanities, and how there is a need for bridging concepts from diverse fields. In this sense, sustainable development has already faced criticism for not clarifying what it really means and attempts to achieve. Two important concepts that need to be clarified to understand sustainable development are,

1. What is to be developed?
2. What is to be sustained?

Again, these two questions can be answered in different ways by various institutions or fields of study, depending on their goals and agendas, which remains a major problem with regards to the use of the term "sustainable development". In addition, sustainable development and sustainability sciences look at enhancing the quality of life rather than mere "survivability" (Pezzey 1992). For example, mainstream economists look at material well-being as the only criterion which has to be sustained, and Brekke defined sustainability as maintaining the net national product (Brekke 1997). Solow argued that it's our obligation to provide capacity or options to improve material well-being to future generations, emphasizing the substitutability of a product by another (Solow 1991). According to him there is nothing wrong with depleting aluminium reserves if our society leaves behind much better material which can replace aluminium reserves (Solow 1991). Herman Daly came out with the concept of steady state growth, with emphasis on human well-being without believing in the substitutability of all capital (Daly 2008). He believes that not all commodities are substitutable, and for example questions the unsustainable use of a large number of fishing fleets with less/no fishing stock. Critical sustainability accepts that not all goods are substitutable, with the major emphasis being on identifying the processes which affect the environment, establishing some safe limit standards or boundaries (Lerch and Nutzinge 2002; Rockström et al. 2009). Sustainability can also be viewed as a political project in addition to an ecological project. Amartya Sen stressed freedom as the ultimate goal of human development, with material well-being increasing human freedom, with an equitable distribution of entitlements and resources leading to future sustainability (Sen 1999).

Sustainability science should consider the complex dynamic system relationships between human and natural systems (Liu et al. 2007; Ostrom 2007; Holling 2005) and the uncertainty associated with them. Research in sustainability must rely on value-laden, normative, and incomplete knowledge, meaning that it is not possible to use traditional research approaches alone. Rather, it is necessary to emphasize a holistic perspective that recognizes the complexity of the systems involved. However, ideas such as complexity and developing a holistic perspective are in themselves complex concepts. Therefore, when tackling complex sustainability issues there also are challenges to make meaningful observations to derive a holistic understanding that can encompass such perspectives (as described in the Chap. 7 and Satanarachchi and Mino 2014).

GPSS-GLI adopts inter-/multi-/trans-disciplinary approaches to understand and deal with those issues. Through guided field exercises academic members of staff attempt to provide a platform where students recognize such challenges by themselves and develop academic, professional, and leadership skills to address them. During the design of the exercises particular attention is paid to attempt to utilize both technical and social scientific research approaches and methodologies. Most of the technical methodologies employed (such as social survey techniques, data analysis techniques) are borrowed from related parent fields of studies, though emphasizing an overarching view of the many issues can help students achieve a top-down approach to tackle the complexities in the field.

3.3 Trans-Boundary Approaches ("Bottom-up")

The complex nature of sustainability issues, and the interdisciplinary and transdisciplinary research (Scholz et al. 2006; Scholz and Tietje 2002; Lang et al. 2012) which is being encouraged to tackle these issues often requires researchers to go beyond boundaries (which can be disciplinary, knowledge, cultural, or even mental boundaries). One way of encouraging a holistic outlook in students is to provide them the grounds for transboundary learning. When attempting to find solutions to a problem, it is important to have respect to and interaction with the other side of boundaries so that new perspectives, understanding, knowledge, wisdom or values can be created and shared by both sides. This aspect becomes especially relevant when students are to explore sustainability issues, often embedded in contexts which are unfamiliar to them. The interdisciplinary and transdisciplinary learning may force them to go beyond their comfort zones and encounter and utilize other sustainability/unsustainability views, research methodologies or techniques. Furthermore, sustainability issues involve several time and spatial frames that require students to be aware of spatial and temporal boundaries. When conducting field research they will naturally be exposed not just to the issues, but also the complexity of interacting with stakeholders with different views, values or priorities. For instance, early field activities in GPSS such as the APIEL (Chiang-Rai unit 2009) explored trans-boundary ideas, particularly related to trans-boundary environmental and social issues in the golden triangle area of Southeast Asia. In these exercises students were exposed to the complexity of the various issues involved and how stakeholders related to those issues, together with the boundaries created by geographical conditions and governing structures. As a conclusion condition and provides reasons for continued adaptation as one of the best ways of facing uncertainties (Holling et al. 2002). The adaptive cycle also allows for a way to understand the temporal scale of a system, though there is nevertheless the need to understand and map resilience on a spatial scale.

However, any discussion on resilience should first start by attempting to define the concept. GPSS-GLI defines resilience as "the capacity of a system, enterprise, or a person to maintain its core purpose and integrity in the face of dramatically changed circumstances" (GPSS-GLI 2015). However, it is important to note that there are a number of other definitions of resilience. For example, the ecological definition tends to tends to focus on the amount of disturbance that a system can take before it switches from one equilibrium regime to another (Gunderson 2002). There are other approaches and definitions, such as those given by the Resilience Assessment Framework, laid out by the Alliance (2010). In 2002, Carpenter suggested the following three possible meanings of resilience, (i) response to disturbance; (ii) capacity to self-organize; and (iii) capacity to learn and adapt (Carpenter et al. 2001). Holling (1986) pointed at the importance of renewal, novelty, innovation and reorganization of a system while extending the concept of resilience to a socio–ecological systems (Holling 1986). However, Walker and later Folke emphasized the additional critical characteristics of a resilient system under a framework called "resilience thinking" , where adaptability and transformation are important features of the system (Folke et al. 2002, 2005). Adaptability has been used in resilience literature, and was defined by Walker as "the capacity of actors in a system to influence resilience" (Walker et al. 2004). These authors defined transformability as a means of defining and creating new stability landscapes by introducing new components and ways of making a living, thereby changing the state variables, and often the scale, that defines the system. The two major components of resilience are,

(1) response to disturbance, or backward looking features of resilience
(2) adaptation and transformation, or forward looking character of resilience

Robustness and innovation are both important for a resilient system. Robustness is important in the context of the present, though innovation becomes important in the context of the future. Flexibility allows a system to adapt to new environments, and will allow it to easily deploy resources where they are needed. An organism or a system has to balance between robustness and transformation in response to changing conditions in the environment. The example of the caterpillar is often cited. Caterpillars need to maintain robustness in terms of functioning and at the same time accommodate the changes in genes required to transform into a butterfly (Ehrlich and Hanski 2004). Similarly, such concepts can be applied to the cultural and social sphere, where competing concerns to stabilize a system and transform it require that a balance between these needs is struck (Ehrlich and Levin 2005). These examples point out the importance of balancing the competing extremes of robustness and transformation, requiring the creation of flexible institutions and processes that can facilitate the achievement of a sustainable and resilient society.

4 Structure of This Book

This book, consists of 10 chapters, summarising some of GPSS-GLI's last four years of educational challenges, as explained earlier. Essentially it is divided into 3 parts, with part 2 describing GFEs, part 3 describing ERs, and then part 4 summarising some field exercise experiences from past and present graduates from the program.

Chapter 2 will explore an integral framework for field work in sustainability science research as well as global leadership education, and show how this framework has been used to conduct one of GPSS-GLI's Global Field Exercise in the Heihe River basin, Northwestern arid China. The framework was developed by modifying Ken Wilber's four-quadrant approach, with the authors' concluding that the use of an integral approach is effective not only for understanding issues of complexity, but also the development and management of environmental leadership education programs.

Chapter 3 introduces how to conduct a system assessment for sustainability science in two steps: an in-depth literature review and field-level survey involving key informant interviews. In employing these two methods, the chapter examines the current challenges to the implementation of Costa Rica's Payments for Environmental Services (PES) for agroforestry and attempts to provide practical recommendations for future improvements. The study was also conducted as part of the Global Field Exercise (GFE)—Costa Rica for the Academic Year 2014, the first time it was implemented. As such, the chapter provides a blueprint of how to conduct an exploratory field study of a given issue in which the participants have limited prior knowledge.

Chapter 4 introduces the Tohoku Unit, an Exercise in Resilience (ER) field exercise conducted in north-eastern Honshu Island in Japan, where students are exposed to the reconstruction process in Otsuchi Town that has followed the Tohoku Earthquake Tsunami of March 2011. An assessment of the field exercise is presented in two levels of depth, (1) an application of field methodologies to identify sustainability issues in the reconstruction process, and (2) the contribution of the fieldwork to the development of student's competencies relevant to sustainability research. The results indicate that the field methodologies utilized contributed to a high number of reconstruction issues being identified. Also, students considered the contribution of the field exercise as "satisfactory" to "effective" in promoting the competencies evaluated. Final remarks highlight the relevance of considering this framework for future improvements in the design of the units by considering the main competencies that they intend to foster in students.

Chapter 5 describes the second of the ER units of the GPSS-GLI, which aims to study the complex, long-lasting and sensitive issues regarding Minamata Disease in Japan. The chapter discusses the efforts that were made by students to understand the sustainability issues involved, and how the unit in the AY 2015 focused on the dissemination of such learning to other sustainability science students and the general public. As a result, group work by the students produced creative outputs that including blog posts, a video, and an educational game, which was the first time that such outputs have been produced by students of a field exercise at GPSS-GLI.

Chapter 6 discusses how the urban planning and design disciplines have repeatedly failed to build sustainable communities that are economically, environmentally, and socially viable and resilient. Sustainability science has the potential to be combined with the fields of urban planning and design, which primarily focus on the physical shape of the city, to develop new methodologies for building sustainable communities. To verify this, the chapter explores potential overlaps by identifying field methodologies and the focus of urban planners and designers, and that of sustainability scientists. Essentially, the authors argue that sustainability science can shift the methodologies used within planning and design towards the use of scientifically oriented methodologies that can help decision-makers create sustainable communities.

Chapter 7 discusses a newly developed methodology to evaluate the sustainability of a human–natural system in a complex dynamic context, which could be useful in conducting sustainability science field research. Evaluating sustainability in human–natural systems requires paying attention to the 'observation process' of these systems to adequately grasp complex dynamics. Failing to do so can result in poor translation of the sustainability/unsustainability patterns. The methodology pays particular attention to the complexities involved in the observation processes that would support reflexive understanding-based sustainability evaluations. The authors discuss the basis of the evaluation methodology and how it can be applied to field research.

Chapter 8 addresses the methodological challenge to examine the sustainability of a target system by applying the community marginalisation framework. This chapter showcases how to apply such framework through a case study in Kamikoani village in Japan. The framework allows the illustration of qualitative changes in rural communities as they experience population decline over a period of time. The findings suggested that there is a clear divergence in the process of functional declines among community groups. Moreover, the methodology applied depicted continuing future declining trajectories of population and community-function for the smallest size of community groups, highlighting the severe problems that are being faced by the entire study area.

Chapter 9 discusses a methodology for envisioning the sustainable development strategies that can be adopted by indigenous communities, through the use of Participatory Mapping and Problem Ranking. Such methods can facilitate the discovery of the local context in sustainability research, based on a review of the present situation in a given community, and an analysis of their recent past. The author concludes that the local conditions inherent to a community need to be systematically analysed to achieve a higher level of participatory development. This can help avoid potential threats to sustainable development assistance to a community through identifying and addressing the local complexity of specific conditions and problems in which the community is embedded in.

Finally, Chap. 10 discusses the methodological implications of selecting sustainability impact assessment criteria. It identifies the wide gap between local impacts experienced and those assessed using globalised unified sustainability impact assessment frameworks. To overcome these challenges, the authors propose the use of a bottom up approach of selecting relevant criteria by making use of a rapid appraisal of sustainability impacts for an initial quick identification of patterns which can then serve as basis for framing the actual sustainability impact assessment methodology.

References

Agrawal, A., Chhatre, A., & Hardin, R. (2008). Changing governance of the world's forests. *Science, 320*(5882), 1460–1462.

Akiyama, T., Li, J., & Onuki, M. (2012). Integral leadership education for sustainable development. *Journal of Integral Theory and Practice, 7*(3), 55–69.

Alliance, R. (2010). Assessing resilience in social-ecological systems: Workbook for practitioners.

Bell, S., & Morse, S. (2008). *Sustainability indicators: measuring the immeasurable?*. UK: Earthscan.

Berkes, F., & Folke, C. (Eds.). (1998). *Linking sociological and ecological systems: management practices and social mechanisms for building resilience*. NY, USA: Cambridge University Press.

Brekke, K. A. (1997). *Economic growth and the environment: On the measurement of income and welfare*. Cheltenham: Edward Elgar.

Carpenter, S., Walker, B., Anderies, J., & Abel, N. (2001). From metaphor to measurement: resilience of what to what? *Ecosystems*, 765–781.

Clark, W. C. (2007). *Sustainability Science: A room of its own*. PNAS.

Clark, W. C., & Dickson, N. M. (2003). Sustainability science: The emerging research program. *Proceedings of the National Academy of Sciences, 100*(14), 8059–8061.

Clayton, A. M., & Radcliffe, N. J. (1996). *Sustainability: A systems approach*. Earthscan.

Daly, H. (2008). *A steady-state economy*. Sustainable Development Commission.

Ehrlich, P., & Hanski, I. (2004). *On the wings of checkerspots: A model system for population biology*. Oxford: Oxford University Press.

Ehrlich, P. R., & Levin, S. A. (2005). The Evolution of Norms. *PLoS Biology, 3*(6), e194.

Espinosa, A., Harnden, R., & Walker, J. (2008). A complexity approach to sustainability–Stafford Beer revisited. *European Journal of Operational Research, 187*(2), 636–651.

Esteban, M., Onuki, M., Ikeda, I., & Akiyama, T. (2015). Reconstruction following the 2011 Tohoku earthquake tsunami: Case study of Otsuchi Town in Iwate prefecture, Japan. In M. Esteban, H. Takagi, & T. Shibayama (Eds.), *Handbook of coastal disaster mitigation for engineers and planners*. Oxford, UK: Butterworth-Heinemann (Elsevier).

Folke, C., Carpenter, S., Elmqvist, T., Gunderson, L., Holling, C. S., & Walker, B. (2002). Resilience and sustainable development: building adaptive capacity in a world of transformations. *AMBIO: A Journal of the Human Environment, 31*(5), 437–440.

Folke, C., Hahn, T., Olsson, P., & Norberg, J. (2005). Adaptive governance of social-ecological systems. *Annual Review of Environment and Resources, 30*, 441–473.

Gibson, C. C., Ostrom, E., & Ahn, T. K. (2000). The concept of scale and the human dimensions of global change: A survey. *Ecological Economics, 32*(2), 217–239.

GPSS-GLI. (2015). Retrieved December 20, 2015, from http://www.sustainability.k.u-tokyo.ac.jp/.

Gunderson, L. H. (2002). *Resilience and the behavior of large-scale systems* (Vol. 60). Island Press.

Hawkes, J. (2001). The fourth Pillar of sustainability. *Culture's essential role in public planning*. Common Ground, Australia.

Holling, C. S., Gunderson, L., & Peterson., G. (2002). Sustainability and panarchies. In L. G. Holling (Ed.), *Panarchy: Understanding transformations in human and natural systems*. Island Press.
Holling, C. S. (2005). *Adaptive environmental assessment and management*, Blackburn Press.
Holling, C. S. (1986). The resilience of terrestrial ecosystems: local surprise and global change. *Sustainable development of the biosphere*, 292–317.
Holling, C. S., Gunderson, L., & Peterson., G. (2002). Sustainability and Panarchies. In L. G. Holling & Panarchy (Eds.), *Understanding Transformations in Human and Natural Systems*. Island Press.
Hopwood, B., Mellor, M., & O'Brien, G. (2005). Sustainable development: Mapping different approaches. *Sustainable Development, 13*(1), 38–52.
Kates, R., William, C., Clark, J., Michael, H., Carlo, J., Ian L., et al. (2000). *Sustainability science*.
Kates, R., Clark, W., Corell, R., Hall, J., Jaeger, C., Lowe, I., McCarthy, J., Schellnhuber, H., Bolin, B., Dickson, N., Faucheux, S., Gallopin, G., Grubler, A., Huntley, B., Jager, J., Jodha, N., Kasperson, R., Mabogunje, A., Matson, P., & Mooney, H. (2001). Sustainability science. *Science, 292*(5517), 641–642.
Komiyama, H., & Takeuchi, K. (2006). Sustainability science: Building a new discipline. *Sustainability Science*, 1–6.
Ladyman, J., Lambert, J., & Wiesner, K. (2013). What is a complex system? *European Journal for Philosophy of Science, 3*, 33–67.
Lang, D. J., Wiek, A., Bergmann, M., Stauffacher, M., Martens, P., Moll, P., & Thomas, C. J. (2012). Transdisciplinary research in sustainability science: Practice, principles, and challenges. *Sustainability Science, 7*(1), 25–43.
Lerch, A., & Nutzinge, H. (2002). Sustainability: Economic Approaches and Ethical Implications. *Journal of Economic and Social Policy*, *6*(2), 2.
Levin, S. A. (2006). Learning to live in a global commons: socioeconomic challenges for a sustainable environment. *Ecological Research, 21*(3), 328–333.
Liu, J., Dietz, T., Carpenter, S. R., Alberti, M., Folke, C., Moran, E., & Taylor, W. W. (2007). Complexity of coupled human and natural systems. *Science, 317*(5844), 1513–1516.
Lozano, R. (2008). Envisioning sustainability three-dimensionally. *Journal of Cleaner Production, 16*(17), 1838–1846.
Mikami, T., Shibayama, T., Esteban, M., & Matsumaru, R. (2012). Field Survey of the 2011 Tohoku earthquake and Tsunami in Miyagi and Fukushima prefectures. *Coastal Engineering Journal (CEJ), 54*(1), 1–26.
Mino, T., & Hanaki, K. (Eds.). (2013). *Environmental leadership capacity building in higher education: Experience and lesson from Asian program for incubation of sustainability*. Japan: Springer.
Neumayer, E. (2003). Weak versus strong sustainability: Exploring the limits of two opposing paradigms. Edward Elgar Publishing.
Olsson, L., Jerneck, A., Thoren, H., Persson, J., & O'Byrne, D. (2015). Why resilience is unappealing to social science: Theoretical and empirical investigations of the scientific use of resilience. *Science Advances*, *1*(4), e1400217.
Ostrom, E. (2007). A general framework for analyzing sustainability of socio-ecological systems. *Proceedings of the Royal Society of London Series B, 274*,19–31.
Page, S. E. (2010). *Diversity and complexity*. NJ: Princeton University Press.
Peter, C., & Swilling, M. (2014). Linking complexity and sustainability theories: Implications for modeling sustainability transitions. *Sustainability, 6*(3), 1594–1622.
Pezzey, J. (1992). Sustainable development concepts. *World, 1*, 45.
Robinson, J. G. (2011). Ethical pluralism, pragmatism, and sustainability in conservation practice. *Biological Conservation, 144*(3), 958–965.
Rockström, J., Steffen, W. L., Noone, K., Persson, Å., Chapin III, F. S., Lambin, E., & Nykvist, B. (2009). Planetary boundaries: Exploring the safe operating space for humanity.

Satanarachchi, N., & Mino, T. (2014). A framework to observe and evaluate the sustainability of human–natural systems in a complex dynamic context. *SpringerPlus, 3*(1), 1–21.
Sen, A. (1999). *Development as Freedom*. Oxford UP: Oxford University Press.
Scholz, R. W., & Tietje, O. (2002). *Embedded case study methods: Integrating quantitative and qualitative knowledge*. Sage.
Scholz, R. W., Lang, D. J., Wiek, A., Walter, A. I., & Stauffacher, M. (2006). Transdisciplinary case studies as a means of sustainability learning: Historical framework and theory. *International Journal of Sustainability in Higher Education, 7*(3), 226–251.
Spangenberg, J. H. (2011). Sustainability science: A review, an analysis and some empirical lessons. *Environmental Conservation*, 275–87.
Solow, R. (1991). *Sustainability: An economist's perspective*. Marine policy centre, Woods Hole Oceanographic Institution, Massachusetts.
Swart, R. J., Raskin, P., & Robinson, J. (2004). The problem of the future: sustainability science and scenario analysis. *Global Environmental Change, 14*(2), 137–146.
Takagi, H., Esteban, M., & Tram, T. T. (2014). Coastal Vulnerabilities in a fast-growing Vietnamese city. In D. T. Nguyen, H. Takagi, & M. Esteban (Eds.), *Coastal disasters and climate change in Vietnam: Engineering and planning perspectives*. Amsterdam: Elsevier.
Walker, B., Holling, C. S., Carpenter, S. R., & Kinzig, A. (2004). Resilience, adaptability and transformability in social–ecological systems. *Ecology and Society, 9*(2), 5.

Part II
Global Field Exercises

Designing Field Exercises with the Integral Approach for Sustainability Science: A Case Study of the Heihe River Basin, China

Ricardo O. San Carlos, Heng Yi Teah, Tomohiro Akiyama and Jia Li

Abstract This chapter introduces one of the GPSS-GLI exercises, the Oasis Unit, which has been conducted annually in northwestern China. The unit aims to equip the students who take it with a wide knowledge base and leadership competency, and is thus strongly field-oriented. The Integral Approach proposed by Ken Wilber is applied when designing this unit, which provides an interdisciplinary and transdisciplinary framework for synthesizing the complex problems of sustainability, as well as allowing the inclusion of methodologies from different academic fields. In the present chapter a practical implementation of the approach is showed and discussed, using as a case study the 2015 Oasis Unit within two contexts—(1) as a holistic framework for addressing sustainability problems, and (2) as a design approach for educational field exercises. Finally, the authors summarize the usefulness of the integral approach in designing field exercises based on the seven year long experience of the University of Tokyo in organizing the Oasis Unit.

Keywords Integral approach · Sustainability science education · Field exercise · Heihe river basin

---

This chapter is an updated version of Akiyama et al. (2010) and Akiyama and Li (2013).

R.O. San Carlos (✉) · H.Y. Teah · T. Akiyama
Graduate Program in Sustainability Science-Global Leadership Initiative,
Graduate School of Frontier Sciences, The University of Tokyo,
5-1-5 Kashiwanoha, Kashiwa, Chiba 277-8563, Japan
e-mail: sancarlos.ricardo@gmail.com

T. Akiyama
e-mail: akiyama@k.u-tokyo.ac.jp

J. Li
Faculty of International Studies and Regional Development,
University of Niigata Prefecture, Niigata Prefecture, Japan

M. Esteban et al. (eds.), *Sustainability Science: Field Methods and Exercises*,
DOI 10.1007/978-3-319-32930-7_2

1 Introduction

The United Nations Decade of Education for Sustainable Development (UNDESD), initiated in 2005, aims to develop and implement educational programs that focus on the three pillars of sustainability, i.e., environment, economy, and society. The Graduate Program in Sustainability Science—Global Leadership Initiative (GPSS-GLI, at the University of Tokyo) was established in line with the Japanese government's initiative to promote sustainability science education in higher education institutions and to nurture global leaders. Since the concept of sustainability science and present-day sustainable development problems feature complex issues, GPSS-GLI has paid considerable attention to interdisciplinary and/or transdisciplinary education. Particularly, through the GLI component of the program the focus is on fostering student leadership through on-site curriculum.

The core of the on-site curriculums are three courses that are known as "Field Exercises", built for the purpose of practical learning. These include Resilience Exercises (RE) and Global Field Exercises (GFE) for master students, and Global Leadership Exercises (GLE) for doctoral students. To guarantee the diversity of education methodologies and educational effects, the field exercise units can be roughly classified as either field-oriented or structure-oriented. In either case, the course consists roughly of four parts: preliminary learning, on-site learning, after-the-fact learning, and a joint task (report preparation and a presentation). Field-oriented exercises develop leadership mainly through student-based group work that consists of reviewing literature, planning on-site research activities, implementing field surveys and presenting the results of the study. Doctoral students who participate in GLE are responsible to coordinate and lead one RE or GFE unit. Akiyama et al. (2012a) mentioned that field-oriented exercises foster global leadership mainly through the execution of field surveys, while the structure-oriented exercises develop leadership through a series of educator-structured programs (i.e., in-class lectures, discussions, and short field trips). Students in GPSS-GLI are diverse in terms of their undergraduate academic disciplines and cultural backgrounds, i.e., 54 graduate students from 20 countries. A holistic framework that emphasizes pluralism and the integration of scientific knowledge, and that is not aiming to establish a single theory (Olsson et al. 2015), is therefore needed in guiding the design of field exercises.

This chapter provides a detailed description of one of the field exercises, the Oasis Unit, which has been conducted annually in the period 2009–2015 in the Heihe River basin in arid northwestern China. The Oasis Unit especially emphasizes on the multidisciplinarity of sustainability science education and the necessity for a holistic view to understand the various dimensions of development issues in the region. Therefore, it makes an effort to incorporate the Integral Approach, adapted from the American philosopher Wilber (2001), into the practice of sustainable development leadership education. The authors introduce the four-quadrant framework of the Integral Approach and two methods of applying the framework in Sect. 2; describe the background of the Oasis unit and the fieldwork carried out through it in Sect. 3; show a case study on 2015 Oasis Unit in Sect. 4; and discuss the applicability of Integral Approach in designing field exercises for sustainability science education in the last section of this chapter.

2 Integral Approach: A Simplified Introduction

2.1 A Four-Quadrant Framework

The Integral Approach aims to incorporate multiple perspectives instead of focusing on specific objects and/or specific systems of objects. Given the complexity of reality, the Integral Approach cuts across fields and brings together existing methodologies into an interdisciplinary and transdisciplinary framework. According to Wilber (2001), all phenomena in the world can be categorized into four groups using a quadrant framework. These quadrants present four ways of viewing the same occurrence in reality from four different perspectives. They are located in the interior and exterior of both individuals and collectives. The exterior aspects are found on the right-hand side, with physical and behavioral aspects in the upper right quadrant, and social systemic aspects in the lower right quadrant. The interior aspects are found on the left-hand side, with intentional, personal, and psychological aspects in the upper left quadrant and cultural aspects (collective values) in the lower left quadrant. Although the four quadrants are ontologically distinct there is nevertheless an interwoven, intimate relationship between them.

Wilber's Integral Approach has received substantial attention and has been applied to a variety of fields, in both academia and practice. In particular, it is gaining attention around the world from researchers and practitioners in the field of sustainability science and environment studies. It has also been previously applied to the fields of development and education. In the field of international development, several international development organizations and non-governmental organizations, including the United Nations Development Program (which started a global leadership development program around HIV/AIDS), are increasingly seeing the advantages of adopting this approach to their projects (Brown 2006). At the local level, this approach has also been used in community development projects such as the one in the San Juan del Gozo community in El Salvador (Hochachka 2008). In the field of education, the approach has also proved useful for curriculum development: see, for example, Gidley and Hampson (2005), Lloyd (2007) and Akiyama et al. (2012a). In the present chapter the authors apply this approach because, first, it provides a holistic framework encompassing a wide knowledge base in the social, economic, cultural, and natural sciences, and second, it leads to the successful implementation of a global leadership education program by providing a comprehensible structure for educational curriculum design. Particularly, the authors will depict and use the quadrant model in two ways: a Quadrivia approach for solving sustainability problems, and a Quadrant approach for the field exercise educational design.

2.2 Quadrivia Approach for Solving Sustainability Problems

Quadrivia, in Latin, means a place where four roads meet. According to the four-quadrant framework (Fig. 1), the Quadrivia approach places a research problem in the center and find solutions from four dimensions, i.e., interior, exterior, individual, and collective. In practice, a research question must first be defined, and then the research methodologies in different fields and disciplines can be cohesively included to answer the question. Structuring a research with this approach is holistic, at least in the interdisciplinary and transdisciplinary context. Figure 1, for example, is a version of Akiyama et al. (2010)'s four-quadrant framework related to water environmental issues. The study covered the objective components, which include hydrological processes and water management systems, and the subjective components, which include personal values and culture.

The authors would like to draw attention on the following studies: Kayane et al. (2006), Kayane (2008a), which pioneered the application of this approach to water environment issues. These authors analyzed the water environment and related changes in Lijiang City, China and Tsuwano Town, Shimane Prefecture, Japan. They argued that, first, the natural environment, especially water, is a common element related to all quadrants, and second, that current environmental problems are often consequences induced by the abnormal development (evolution) of the lower right quadrant, i.e., rapid technological innovation in the 20th century. The authors refer to such work because the main topic of the Oasis Unit is water scarcity and

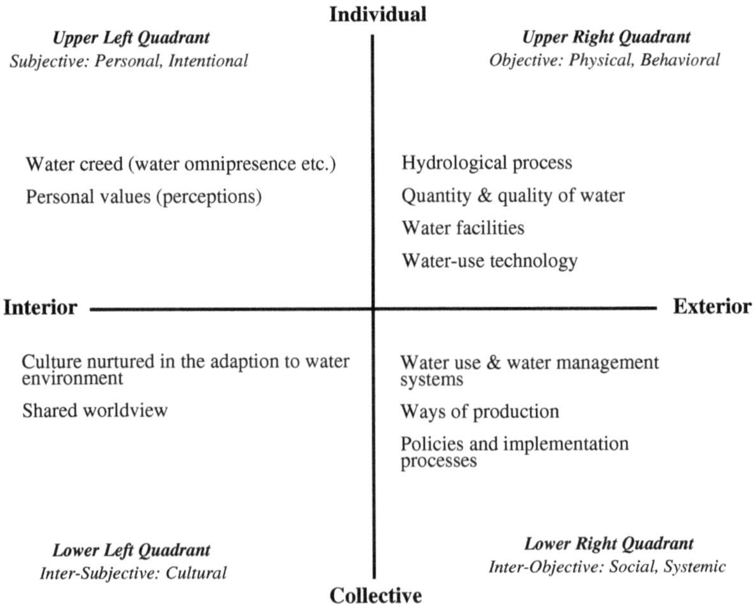

Fig. 1 Akiyama et al. (2010)'s four-quadrant framework related to water environmental issues

water resource management in arid regions. The framework outlined in Kayane et al. (2006) and Kayane (2008b) was further developed in Akiyama et al. (2012b).

The framework used in the present chapter allocates perspectives on water environmental issues into each of the four quadrants. It draws on the philosophy of sustainability science and environment studies, with its emphasis on empirical research methods (quantitative and scientific), as well as philosophy, to encompass inter-subjective and subjective modes of inquiry (qualitative and social-value oriented). The benefit of this framework, although requiring further research, is profound. It offers a common foundation to view the various perspectives on the complexity of water environmental issues. In other words, it incorporates knowledge and methodologies from multiple disciplines.

2.3 *Quadrant Approach in the Development of Field Exercise Education*

The integral approach presents some advantages in tackling sustainability or environmental issues, but it is not the only way that the framework can be utilized within the Oasis Unit. Since the Oasis Unit is part of a research and educational program, it is imperative that its design considers not only the methodologies to address the object of study (Quadrivia) but also how it will impact the learning and growth of the individuals/students involved (Quadrant). Thus, in this section the authors will discuss the characteristics of the integral approach for the design of a field exercise like the Oasis Unit. For this discussion the authors will focus on the application of the integral approach to the conceptual design of a field-based and problem-oriented educational program, while addressing its advantages, limitations, opportunities and challenges. Finally, the authors will reflect on the case study of the Oasis unit carried out in summer of 2015.

In Fig. 2, "I" (or "we") refers to the participant(s) of the field exercise. The framework first requires fostering self-development through personal learning as well as group work, collaboration and communication with the different stakeholders. Then, it also requires all participants to achieve common conclusions, as well as to accommodate their individual views. Therefore, the field exercise should provide enough time and resources for the participants to reach a consensus and to set shared clear goals from the beginning, while also allowing for individual points of view.

Considering the design of a field exercise, the educational goals of the program should generally be stated beforehand. The set of competencies expected as a result of the field exercise could constitute the main outcome from the point of view of an individual's learning and growth as a sustainability researcher. The issue at hand is that the necessary competencies for sustainability scholars are still matter of debate. Although there are some important studies regarding the competencies in sustainability science (Wiek et al. 2011), these are not yet completely accepted across all sustainability programs and institutions.

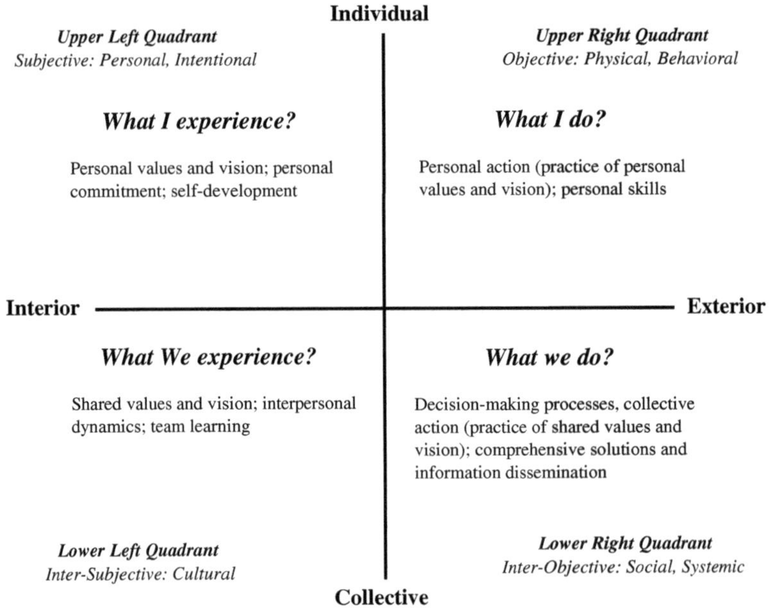

Fig. 2 An integral framework for designing the education process of field exercises

An alternative to the discussion of the competencies necessary for sustainability science could be found in the application of the integral approach. Instead of focusing on the expected outcomes, the design could be focused on the experience of the individual during the field exercise. According to the integral approach, every experience can be understood by applying the 4 irreducible dimensions described by the quadrants. In this way, what is relevant when designing a field exercise is to provide individuals with experiences that allow them to reach desired levels of depth and complexity within each quadrant. Then, the focus is not on identifying a specific set of competencies to develop, but on designing different experience channels for the individual that relate to sustainability issues (Esbjörn-Hargens 2005). In turn this will translate into deeper and more complex levels of learning for the individual in all four dimensions of the quadrant.

3 Applying the Integral Approach to Field Exercises

3.1 Oasis Unit in Northwestern Arid China

The Heihe River basin in arid northwestern China is an excellent area for fieldwork participants to consider how sustainable development could be achieved in dryland regions under severe water resources constraints. The University of

Tokyo's Asian Program for Incubation of Environmental Leaders (APIEL), later succeeded by GPSS-GLI, chose this river basin specifically as a target area for field exercise for the two reasons detailed below.

First, the sustainable development of dryland regions is one of the major challenges facing the planet nowadays. It is associated with water security as well as food security around the world. Today, irrigated agricultural land makes up less than one-fifth of the total cultivated area in the world, but produces about two-fifths of the world's food (Pirages 2000). Irrigation farming, to a great extent, contributed to the increase in food production in the 20th century, and continues to support the planet's growing population. However, food production -relying on the "irrigation miracle"—has a significant impact on water resources. Agricultural water use, including irrigation, accounts for about 70 % of global water usage (Shiklomanov 2000). In dryland regions, large-scale development of irrigation farming has induced dramatic increases in water demand. Consequently, it often results in the stoppage of river flows, dry-up of lakes, decline of groundwater tables and other related ecosystem degradation.

Second, the Heihe River basin, the second largest inland river in China, provides many topics for the study of sustainable development in dryland regions. For the case of the Oasis Unit these include watershed management, water-saving policies (decision-making processes, implementation and assessment), as well as environmental degradation and recovery. In the Heihe River basin, historically, people living in the middle and lower reaches have adopted different ways of production, namely irrigation farming (settled culture) in the middle reaches and nomadic husbandry in the lower reaches. Intensive agricultural practices in the middle reaches since the 1950s have resulted in a dramatic degradation of the environment downstream (in the lower reaches). Conflicts over water use between people living in the middle reaches and those living in the lower reaches dates back to at least 200 years ago (Inoue 2007). However, these conflicts have never been as fierce as today. The intensive exploitation of water resources in the middle reaches has cut the water flow to the lower reaches. By 2002, more than 30 tributaries of the Heihe River basin had dried up. In the lower reaches, two terminal lakes dried up in 1961 and 1992. Riparian vegetation degraded. Salinization and desertification intensified. The desertification is attracting substantial attention nationwide and is thought to be the origin of dust storms in the spring.

In recent years, a range of environmental conservation activities has been carried out in the river basin, particularly around Zhangye, a city in the middle reaches of the Heihe River basin. The main purpose of environmental conservation activities is to preserve the environment in the lower reaches. At the core of those activities is the Integrated Water Resources Management Plan of the Heihe River basin, promulgated by the Chinese state council in 2001. This plan states that "when the water from the upstream discharge reaches 1.58 billion m^3/a, Zhangye, located in the middle reaches of the Heihe River basin, has to increase discharge 0.225 billion m^3/a to the lower basin, which means 0.95 billion m^3/a should be released to the downstream" (Fang et al. 2007). In other words, the central government requires the city of Zhangye to reduce water consumption by means of an administrative order. Since 2001, Zhangye has been repeatedly selected as an experimental site for pilot programs for water resource management. In early 2002, the Ministry of Water Resources of China initiated an experimental project for establishing a water-saving society in the middle reaches at Zhangye. The project was set to save water and increase water use efficiency, mainly in two ways: (1) by building concrete irrigation channels using government funds; (2) by introducing market mechanisms. The policies include the introduction of meters to charge for irrigation water based on the amount used, and the introduction of a water use rights system with tradable water quotas. At the same time, in Ejina (in the lower reaches) a relocation policy has been implemented, as overgrazing in the area was considered one of the reasons behind the environmental degradation.

3.2 Integral Design of the Oasis Unit

Applying the Integral Approach to global leadership education is an evolving process that is far from complete. There are external constraints, such as those on human resources, finances and time, which prevent the ideal development of a program. In addition, students, who have come through a relatively narrow educative system, do not always know how to respond to a new, holistic way of learning. Therefore, during the design the Oasis Unit educators need to focus on an integral knowledge base as well as integral practices.

Started in 2009, the Oasis Unit has taken place once a year up to the time this chapter was written. With integral thinking as the general framework for program design, its organizers have extended the content of the field exercise to incorporate perspectives related to environmental issues, and provided more experiences so that students could develop practical skills. The field exercise is jointly organized by GPSS-GLI, The University of Tokyo, and the Cold and the Arid Regions Environment and Engineering Research Institute (CAREERI), of the Chinese Academy of Sciences. Students who join the field exercise belong to either of these institutes, and come from different countries and undergraduate backgrounds. To provide the students with multi-disciplinary knowledge and multiple views about local environmental problems, faculty members from different academic fields as well as local stakeholders are typically involved in the different stages of the fieldwork. Its organizers have established close relationships with the local water authority to move beyond the boundaries between academia and stakeholders, as well as to let students know that they are tackling real-world problems. Students are required to make policy recommendations and deliver this information to the local water authority. In addition, in 2011 and 2012, the collaboration was strengthened by working with several other institutions from both Japan and China. Organizers have thus held international symposia in Japan and in China to build a platform for students to hear fresh voices from academia beyond faculty members, government officials, and business people.

Fig. 3 An integral organizational framework for Oasis Unit (Akiyama and Li (2013))

Figure 3 is an overview of the organizational framework used in the field exercise. Note that students are the leading players. Educators simply created the space for students to see real-world environmental problems and to realize their own development. In Fig. 3, environmental issues (*Issues addressed*) are the research topics covered by the students; *methodologies* are those adopted by the students; and *competencies* are the capabilities and/or skills that students are expected to have after participating in the field exercise.

Problem-solving based learning is the core concept of the course design. It reveals related issues, brings together the necessary research methodologies, and consequently improves participants' competence to become global leaders in the future. The main objective of the field exercise is to enhance the students' practical skills through solving specific environmental problems in the real world. Issues in each quadrant have different perspectives for the same environmental problem: the sustainable development of the Heihe River basin, which is facing severe water shortages. The issues are interwoven. To provide comprehensive solutions for multiple issues, different methodologies from diverse fields are required. Though they cut across quadrants, natural science methods, including experiments and quantitative modeling, are mostly required to tackle the issues in the upper right quadrant. For the lower right quadrant, social science methods are mostly required. In the case of the two left-side quadrants, humanity-based, hermeneutic methods are mostly required.

The competencies that were identified in Fig. 3 were not intentionally selected by the organizers. They developed naturally in the process of participating in fieldwork, in particular through group work (which requires the students to listen to, understand, and assimilate different ideas). Students need to find common research interests and decide on common research topics, as well as adapt to change, and finally to solve the problems. In addition, competencies spill over to other quadrants. For example, good communication skills may foster students' understanding of the variety of stakeholder concerns, to create a shared vision, and to integrate methodologies and fields to find comprehensive solutions.

4 Case Study of 2015 Oasis Unit

The 2015 Oasis unit was started in May, with the fieldwork being conducted from the 1st to 14th August in Lanzhou and Zhangye. Members of the unit consisted of one doctoral student, four master students, and three faculty members. Collaborative members were students and faculty from Sophia University, Tokyo University of Agriculture and Technology, and CAREERI, 53 members in total.

4.1 Designing Sustainability Assessment in Quadrivia View

Before defining the research theme, students were advised to review the 25 top-cited and the 25 latest literatures about the region, as outlined in the Web of Science platform. The keywords "Heihe River basin" and "Zhangye" were used as filters. The results encompassed literature from a wide range of disciplines, including hydrology, remote sensing, decision-making modeling, land-use and cover-change, water management, economics, and policy analysis. The literature provided rich reference to formulate an integrated study. Students were then encouraged to propose research projects under the umbrella of sustainable development. A supplement of project reports written by previous Oasis Units was provided as references. Proposals were made mainly based on students' personal interest or their initial training (i.e., environmental science and engineering, civil engineering, economy, and literature study).

A research theme, "A Planetary Boundary based Framework for Sustainability Assessment of Heihe Middle Reaches", was finally agreed upon by all members. It used the point of view of Planetary Boundaries (Rockstrom et al. 2009; Steffen et al. 2015), defining sustainability as the development of human activities that do not transgress ecological limitations, for carrying out the Oasis regional sustainability assessment. First, the nine global boundaries proposed in Planetary Boundary (Steffen et al. 2015) were reduced to five ecosystem boundaries, i.e., freshwater use, biogeochemical flow, atmospheric aerosol loading, novel entities, and land-system change, based on the existence of regional thresholds. The measurable indicators and the values for each boundary were then downscaled to the local level to reflect the ecosystem in the region (Cole et al. 2014; Dearing et al. 2014). Environmental data for the study area, the middle reaches of Heihe River basin, was collected and analyzed. The result showed that the freshwater use of the region is at high-risk level, and that it critically outweighed other boundaries.

In order to provide useful recommendation to the local government, the unit performed an attempt to simulate how economic activity contributed to each boundary. First, the GDP contribution of each industrial sector and their environmental loading were reviewed based on available data. Then, an estimation model could be built. An assessment of the future sustainability status was made possible by using the expected GDP growth in the local government's five-year-plan. The model could thus contribute to assessing and comparing alternative policies, for example, the impact of transforming economic structure from primary to tertiary industry.

In the view of the four-quadrant framework, the preliminary assessments mentioned above first focused on the upper right quadrant (i.e., aspects of exterior and individual) in studying the environmental facts of the Heihe River basin. Then, it moved downward to the bottom right quadrant (i.e., aspects of exterior and collective) in an expansion to management policy. If the research theme is placed at the center of the integral framework, the knowledge from the left quadrants, or "interior" aspects, regardless of individual or collective perspectives, were missing. In other words, the unit had been focusing only on the holistic view of interdisciplinary research on "exterior" aspects, which was conventionally quantifiable through the methodologies of traditional research disciplines. To bring in the elements of transdisciplinary research, the question to be answered was how the subjective view of individual and/or socio-cultural factors could respond to the proposed sustainability assessment.

The underlying concept of Planetary Boundary is based on the premise that humanity could face a high-risk of various types of ecological collapses if development transgresses nature's capacity to sustain it. The quantitative definition of the level of high-risk or boundary value must involve a normative value of human judgment (Dearing et al. 2014) that could not be solely answered in scientific terms. An example question would be, what percentage of land should be preserved as natural landscape to maintain the health of a regional ecosystem. Although we know the function of natural landscape in protecting biodiversity and regulating natural cycles, we are uncertain in quantifying such effect. The setting of boundary value must depend on collective local cultural judgments and values. Therefore, during the fieldwork in Zhangye a structured interview survey was conducted to examine the public perception on the sustainability of local ecosystem. Individual opinion was explored in two ways: how people perceived the changes in the environment with relation to the impact on their livelihood, and the urgency of different environmental risks. To do so, the members of the unit visited 21 villages and interviewed 58 residents along the middle reaches of Heihe River basin.

The results showed that a majority (79 %) of interviewees were concerned about the water scarcity impacting their livelihood, but did not agree regarding further deterioration on freshwater availability (22 % considered that reduction of river water would not take place in the future). One of the reasons given was that they had become aware of the water conservation effort by the government, and had a strong faith in the results of such policy. Examining the collection of individual interviews enriched the research with a sense of cultural or shared views by local community.

In short, the design of sustainability research of Oasis 2015 achieved a complete exploration of knowledge from all four quadrants, which is aligned with the Quadrivia approach. Its participants managed to contribute to a novel aspect to the Planetary Boundary, i.e., exploring the differences between people's perception and scientific determination of one regional ecological boundary.

4.2 Designing the Educational Program in Quadrants View

Akiyama and Li (2013) reviewed some of the learning outcomes from earlier Oasis Unit exercises carried out in the period from 2009 to 2012. These learning outcomes are described as competencies developed by students and are categorized depending on the quadrant most representative of each given competency. In brief, the results of Akiyama and Li (2013)'s study regarding the competencies developed by student are shown in Fig. 4.

Following the results presented by Akiyama and Li (2013), the authors have complemented the design analysis of the field exercise with the results obtained from the 2015 Oasis Unit. In terms of design analysis, the results are analyzed based on the activities carried out by the student during the 2015 Oasis Unit, and compared to the competencies found in 2013. The Oasis Unit can be decomposed into three chronological stages: pre-fieldwork (preliminary learning), fieldwork (on-site learning), and post-fieldwork (after-the-fact learning and joint task). At the moment of writing this chapter only the first two stages had been completed in their entirety, as shown in Table 1. Additionally, more time would be required to observe the development of competencies in the participants of the 2015 Oasis Unit. For these reasons, the results regarding the analysis of the field exercise design are presented as a preliminary case study in the present chapter. In spite of this limitation, the authors believe that the observations from the completion of the first two stages and the advancements on the third stage already present relevant information for the discussion of the design of the field exercise in 2015.

From Fig. 5 it is possible to appreciate how the activities carried out during Oasis Unit 2015 correlate well with the competencies developed by students from previous versions of this field exercise. It is important to bear in mind that the activities and competencies are expedientially classified into independent quadrants, but in reality they are interwoven and it is to be expected that activities mostly associated to one quadrant could have an effect on developing a

Upper Left Quadrant *Subjective: Personal, Intentional*	Upper Right Quadrant *Objective: Physical, Behavioral*
▪ Finding personal vision ▪ Capacity to engage in self-reflection and introspection ▪ Increased self-awareness and emotional intelligence ▪ Increasing self-esteem ▪ Self-confidence and accountability	▪ Technical skills for independent research ▪ Facilitating communication ▪ Negotiating ▪ Decision making
▪ Creating shared vision ▪ Valuing different perspectives ▪ Communication, listening, and interpersonal skills ▪ Observing and understanding the dynamics of different stakeholders ▪ Building trust	▪ Problem solving ▪ Building a network with resource persons ▪ Inclusion, listening and using all available ideas and skills ▪ Proactive information dissemination ▪ Bringing local voices into decision making
Lower Left Quadrant *Inter-Subjective: Cultural*	Lower Right Quadrant *Inter-Objective: Social, Systemic*

Fig. 4 Learning outcomes obtained from the Oasis Unit in the period 2009–2012

Table 1 The three stages of Oasis Unit 2015

Pre-fieldwork	Fieldwork	Post-Fieldwork
• Literature review • Group meetings • Presentation of research interest • Research proposal • Survey design • Environmental data collection and analysis • Coordination meetings with collaborating universities • Selection of group research	• Visit to research center • Survey tests • Survey re-design • Interview design for city officials • Day-to-day individual tasks • Training of local students • Visit to 21 local villages • Survey to 58 villagers • Interview with city officials • Visit to water management related facilities • Visit to historical/cultural locations in the area • Research progress presentations	• Post-fieldwork meetings • Transcription of recordings • Transcription of notes • Data analysis • Paper/report writing • Final presentation to collaborating universities and local government

competency mostly associated with another one. Further research is still required to verify the type of competencies developed by the students involved in the 2015 Oasis Unit, but at this point it is possible to demonstrate that the design of the unit allows for activities that enhance the student's experience in all four dimensions of the quadrants (Fig. 5). It is also interesting to note that for each quadrant there are activities that cover all three stages of the field exercise.

Upper Left Quadrant *Subjective: Personal, Intentional*		Upper Right Quadrant *Objective: Physical, Behavioral*	
Activities	Competencies	Activities	Competencies
▪ Literature review ▪ Research proposal ▪ Day-to-day individual tasks ▪ Paper/Report writing	▪ Finding personal vision ▪ Capacity to engage in self-reflection and introspection ▪ Increased self-awareness and emotional intelligence ▪ Increasing self-esteem ▪ Self-confidence and accountability	▪ Presentation of research interest ▪ Training of local students ▪ Visit to 21 local villages ▪ Survey to 58 villagers ▪ Research progress presentations ▪ Interview with city officials ▪ Transcription of recordings ▪ Transcription of notes ▪ Data analysis	▪ Technical skills for independent research ▪ Facilitating communication ▪ Negotiating ▪ Decision making
▪ Selection of group research ▪ Coordination meetings with collaborating universities ▪ Survey tests ▪ Visit to historical and cultural locations in the area ▪ Visit to research center ▪ Visit to water management related facilities ▪ Post-fieldwork meetings	▪ Creating shared vision ▪ Valuing different perspectives ▪ Communication, listening, and interpersonal skills ▪ Observing and understanding the dynamics of different stakeholders ▪ Building trust	▪ Survey design ▪ Group meetings ▪ Environmental data collection and analysis ▪ Survey re-design ▪ Interview design for city officials ▪ Final presentation to collaborating universities	▪ Problem solving ▪ Building a network with resource persons ▪ Inclusion, listening and using all available ideas and skills ▪ Proactive information dissemination ▪ Bringing local voices into decision making
Activities	Competencies	Activities	Competencies
Lower Left Quadrant *Inter-Subjective: Cultural*		Lower Right Quadrant *Inter-Objective: Social, Systemic*	

Fig. 5 A summary of activities in Oasis Unit 2015 and competencies found in 2009–2012 (An et al. 2011)

Since the field exercise is designed to address the issue of sustainability and it provides activities to ensure experiences in all four dimensions of the quadrant, it is expected that the competencies developed would be (at least in part, if not entirely) those needed for sustainability researchers. This is an interesting thought that should be explored in future research to contribute to the discussion regarding the competencies required in sustainability science and the overall design of educational programs in this field.

5 Concluding Remarks: Integral Approach and Design of Oasis Unit

The Oasis Unit has been in constant development since its conception. Although the educational outcomes seem to be adequate and satisfactory so far, it is necessary to keep a critical perspective on the design of the field exercise to maintain this tendency and in order to ensure future improvement.

The strength of the integral approach lies in the fact that it provides a structured framework for the transdisciplinary research of complex sustainability issues, and decomposes the experiences of individuals into their four irreducible dimensions. In this way, it is possible for the designer(s) of the field exercise to understand the dimensions that the planned activities will affect the most, and consequently design an exercise that would allow the application of a range of diverse research methodologies and learning experiences with increasing depth and complexity in each dimension.

Because of its holistic nature, this approach can be especially challenging to implement or to be put into practice, since it does not provide clear steps for the design of a field exercise. Also, although it enables the designer(s) to check whether the planned activities could provide experiences in the four dimensions of the quadrant, it does not ensure that the outcome will be the same for all individuals. Given that the design of Oasis Unit considers the participation of students from diverse cultural and academic backgrounds, and despite the fact that they are all exposed to similar activities during the field exercise, the outcomes could vary greatly from student to student depending on their own personalities and idiosyncrasies.

The participation of other institutions during the field exercise is an opportunity for students to expand the range of their experiences, adding depth and complexity to the original design of the field exercise. Nonetheless, if the collaborating institutions are following a different approach this can result in a challenge to coordinate common activities in a way that does not hinder the original activities planned for the students. The coordination with other institutions and administrative tasks should not get in the way of the overall experience of the field exercise, unless of course they were planned as part of it from the beginning.

In conclusion, the true value of using an integral framework is that it is able to (1) map the research issue and methodologies in a holistic manner that is unique to sustainability science, and (2) design the educational program of a given field exercise. Compared to a conventional single problem focus fieldwork, it provides the cohesion for interdisciplinary scholars and transdisciplinary actors to come together in an integrated framework. The four quadrants are irreducible. If a simplification of the situation was attempted by leaving out one or more perspectives, a fundamental aspect of the integral whole would be lost and the ability to understand and address it would be compromised.

This chapter supports most of the findings of the previous work carried out by Akiyama and Li (2013), though it adds new depth to the understanding of the application of Integral Approach to the design of field exercises. By analyzing the relation between activities and competences developed by students it is possible to understand the effect that experiences in each dimension of the quadrants have on their educational growth as sustainability researchers. Thus, the Integral Approach helps in the initial design of a field exercise as well as to when thinking about improvements of it, and it can significantly contribute to the discussion on relevant competences for sustainability researchers. The authors therefore suggest that field exercise practitioners apply the Integral Approach prior and after any research endeavor, for initial scanning and understanding of a given research problem and the design of activities and experiences on the field.

Acknowledgements A number of people have been part of the discussion for developing the field exercise unit conducted in the Heihe River basin. Especially highly acknowledged are the initiatives by Prof. Xin Li (Cold and Arid Regions Environment and Engineering Research Institute, Chinese Academy of Sciences), Prof. Mingguo Ma (Southwest University), Prof. Guangwei Huang (Sophia University), Prof. Takashi Mino, Prof. Eiji Yamaji, and Prof. Tomochika Tokunaga (The University of Tokyo) for the start-up of the field exercise unit. However, the authors take sole responsibility for any errors and the interpretation provided in this chapter.

References

Akiyama, T., & Li, J. (2013). Environmental leadership education for tackling water environmental issues in arid regions. *Environmental leadership capacity building in higher education*, (pp. 81–92). Springer.

Akiyama, T., Li, J., Kubota, J., Konagaya, Y. & Watanabe, M. (2012b). Perspectives on sustainability assessment: An integral approach to historical changes in social systems and water environment in the Ili River basin of Central Eurasia, 1900–2008. *World Futures, 68*(8): 595–627.

Akiyama, T., Li, J., & Onuki, M. (2012a). Integral leadership education for sustinable development. *Journal of Integral Theory and Practice, 7*(3): 55–69.

Akiyama, T., Li, J., Tokunaga, T., Onuki, M., An, K. J., Hoshiko, T., & Ikeda, I. (2010). Integral approach to environmental leadership education: An exploration in the Heihe River basin, Northwestern China. In: *Proceedings of the 8th international symposium on Southeast Asian water environment*, pp 40–49.

An, K., Akiyama, T., Kim, J., Hoshiko, T., & Furumai, H. (2011). The influence of field-oriented environmental education on leadership development. *Procedia Social and Behavioral Sciences, 15*, 1271–1275.

Brown, B. (2006). The use of an integral approach by UNDP's HIV/AIDS group as part of their global response to the HIV/AIDS epidemic. Retrieved 9 May 2010, from http://www.integralworld.net/pdf/Brown.pdf.

Cole, M. J., Bailey, R. M., & New, M. G. (2014). Tracking sustainable development with a national barometer for South Africa using a downscaled "safe and just space" framework. *Proceedings of the National Academy of Sciences, 111*(42), E4399–E4408.
Dearing, J. A., Wang, R., Zhang, K., Dyke, J. G., Haberl, H., Hossain, M. S., et al. (2014). Safe and just operating spaces for regional social-ecological systems. *Global Environmental Change, 28*, 227–238.
Esbjörn-Hargens, S. (2005). Integral ecology: The what, who, and how of environmental phenomena. *World futures, 61*(1–2), 5–49.
Fang, C.-L., Bao, C., & Huang, J.-C. (2007). Management implications to water resources constraint force on socio-economic system in rapid urbanization: A case study of the Hexi Corridor, NW China. *Water Resources Management, 21*(9), 1613–1633.
Gidley, J. M., & Hampson, G. P. (2005). The evolution of futures in school education. *Futures, 37*(4), 255–271.
Hochachka, G. (2008). Case studies in integral approaches in international development. *Journal of Integral Theory and Practice, 3*(2), 58–108.
Inoue, M. (2007). Shindaiyouzeinenkan ni okeru Kokuga no danryu to kokugakinsuiseido ni tsuite. *Oasis chiikishi ronsou (in Japanese)*. M. Inoue, Y. Kato & K. Moriya, 173–191.
Kayane, I. (2008a). Reikoukojou to Tsuwano no hikaku kara kangaeta tourism no yakuwari (in Japanese). *Database of international center for Chinese studies*, Aichi: Aichi University.
Kayane, I. (2008b). Environmental problems in modern china: Issues and outlook. *New challenges and perspectives of modern Chinese studies*, 265–285.
Kayane, I., Miyazawa, T., & Zhu, A. (2006). Reikoukojou no mizu to shakai (in Japanese). *Water Sciences, 291*, 41–72.
Lloyd, D. (2007). An integral approach to learning. *The International Journal of Environmental, Cultural Economic and Social Sustainability, 2*, 27–33.
Olsson, L., Jerneck, A., Thoren, H., Persson, J., & O'Byrne, D. (2015). *Why resilience is unappealing to social science: Theoretical and empirical investigations of the scientific use of resilience*.
Pirages, D. (2000). Pillar of sand: Can the irrigation miracle last? *Perspectives on Political Science, 29*(2), 121.
Rockstrom, J., Steffen, W., Noone, K., Persson, A., Chapin, F. S., Lambin, E. F., et al. (2009). A safe operating space for humanity. *Nature, 461*(7263), 472–475.
Shiklomanov, I. A. (2000). Appraisal and assessment of world water resources. *Water International, 25*(1), 11–32.
Steffen, W., Richardson, K., Rockström, J., Cornell, S. E., Fetzer, I., Bennett, E. M., et al. (2015). Planetary boundaries: Guiding human development on a changing planet. *Science, 347*(6223).
Wiek, A., Withycombe, L., & Redman, C. (2011). Key competencies in sustainability: A reference framework for academic program development. *Sustainability Science, 6*(2), 203–218.
Wilber, K. (2001). *A theory of everything: An integral vision for business, politics, science, and spirituality*. Shambhala publications.

1 Introduction

This chapter aims to understand how the Costa Rican Payments for Environmental Services (PES) attained its objectives and at the same time examine current issues and challenges faced by PES facilitators and implementers. The data was collected through literature reviews and field-level interview surveys with key stakeholders. The methodology employed can be seen as a sort of preliminary assessment of a given issue in sustainability science, where a group of students can set out to make an assessment of the current state of a certain system by conducting an in-depth literature survey and preliminary field study that interviews key stakeholders. As such, the present chapter provides a blueprint of how to conduct an exploratory field study of a given issue in which the participants (including faculty members) have limited prior knowledge. Sustainability scientists and practitioners often must undertake such exercises in order to clarify whether a given system is sustainable or not and the contribution it makes to literature in general and sustainability science in particular. Such exercises and activities are typical in the GPSS-GLI (see Chap. 1), with the present chapter describing a sample the outcome of the first year in which a GFE exercise was run, GFE Costa Rica.

From having one of the world's highest deforestation rates, Costa Rica became the first country in the developing world to successfully reverse deforestation. According to FONAFIFO (2013) the country's forest cover accounted for 53 % of its continental territory in 2013, compared to only 21 % in 1987. This success has been partly credited to the countrywide program on PES implemented in 1997, in addition to various forestry incentives implemented since the late 1970s (Pagiola 2006).

The PES program provides cash remuneration to landowners who ensure the sustainable generation of ecosystem services through the adoption of specific land use modalities (FONAFIFO 2013). There are four explicitly recognized environmental services in PES: (i) provision of scenic beauty; (ii) biodiversity protection; (iii) mitigation of greenhouse gas emissions; and (iv) protection of water for rural and urban use, and for hydroelectric plants. From 1997 to 2000 payments were given out for forest protection, reforestation, and sustainable forest management, which are referred to as the first generation PES modalities. Land uses included after the year 2000 are classified as second-generation modalities.

The study examined whether agroforestry systems, implemented in 2003 as one of the second-generation modalities, satisfies the goal of PES in ensuring sustainable ecosystem service generation. The success of PES for agroforestry was investigated from the perspective of stakeholders from the government, academia and NGOs who were actively involved in its implementation. Unlike earlier land use modalities that provide payments by hectare, PES for agroforestry pays according to the number of trees planted. However, this scheme is deemed problematic by stakeholders due to its minimal regard on the quality of agroforestry farms and its primary focus on increasing tree cover, which does not necessarily guarantee the optimal provision of ecosystem services.

Focusing on the implementation structure of PES facilitated a methodical comparison of policy objectives and actual service delivery. Understanding the goals, strategies, activities and contacts of the network of actors involved in service delivery enabled the authors to examine whether formal policy objectives were attained and at the same time identify opportunities for the program to achieve more than its intended policy results.

The chapter starts by describing the methodology, and then goes on to describe the theoretical framework of Costa Rican PES and discuss its implementation on agroforestry. Then, it examines the implementation structure of the program, and analyzes interview and field observation data to identify key aspects that can be improved to enhance the quality of ecosystem services in agroforestry farms. Finally, the chapter concludes with implications of the study for the future development of PES for agroforestry.

2 Methodology

The study was conducted as part of the Global Field Exercise (GFE)—Costa Rica Unit of the Graduate Program in Sustainability Science-Global Leadership Initiative (GPSS-GLI) for the Academic Year 2014 (see Chap. 1 in the present book).

2.1 Theoretical Framework: Sustainability Science, PES, and Agroforestry

2.1.1 Sustainability Science

Over the years sustainability science has developed a core research agenda focused on problem solving to facilitate societal transition toward sustainable development (Clark 2006). However, providing a solution to sustainability problems is complicated by the fact that they tend to be elusive and intractable (Rittel and Webber 1973). Since such problems occur in a social context—often involving multiple stakeholders who can have radically different understanding of the real problem (Horn and Weber 2007)—coming up with definitive solution is typically difficult. Nevertheless, Batie (2008) noted that stakeholder engagement is important in sustainability science research since they can help frame the problem, determine appropriate goals and most importantly they are in the position to implement the desired change.

In recognition of these fundamental concepts in sustainability science research, the authors, throughout the study, endeavoured to diligently engage the most important stakeholders of PES from various facets of implementation to come up with a holistic view of the actual situation. Interviews were kept as open as possible to avoid influencing stakeholder opinions with the authors' own assumptions. In this way, the different values and perceptions of the stakeholders were spontaneously brought out during the discussions that ensued.

2.1.2 Payment for Ecosystem Services

Wunder (2005) defines PES as a voluntary transaction where a well-defined environmental service (or land use to secure the service) is being "bought" by a service buyer (minimum of one) from a service provider (minimum of one) if and only if the provider secures provision (conditionality). This definition underpins the theory that PES is an effective way to induce conservation by lessening the conflict between conservation and local welfare in compensating those who bear its cost (Pfaff et al. 2008).

However, transaction costs remain a major bottleneck for PES implementation (Wunder 2008a, b). Transaction costs can be incurred from information dissemination, contract negotiation, assessment of baseline environmental service, etc. Additional costs from monitoring, enforcement, as well as administration, are further accrued during the operational phase. In the Costa Rican PES, transaction costs are mostly borne by the national government, who acts as buyer on behalf of environmental service users.

Nonetheless, PES participants still have to bear some of the start-up costs, which sometimes hinder participation. One of the prerequisites of PES is conditionality; as proof of ownership and a minimum land area is required from PES participants. However, these qualifications have inadvertently limited the accessibility of PES to smallholders who cannot meet the minimum area requirement or possess legal proof of land ownership. PES for agroforestry was later introduced, with a payment scheme based on number of trees instead of land area. Special consideration for informal land ownership was also implemented alongside this reform. In principle, PES for agroforestry addresses the shortcomings of earlier PES modalities by introducing reforms in the qualification requirements to make it more accessible to smallholders.

2.1.3 Agroforestry

Agricultural lands are increasingly recognized as a valuable source of various environmental services (Clay 2004). Although in the past agriculture was largely viewed as harmful to biodiversity and soil conservation, more recent studies suggest that with the adoption of appropriate land use systems such as agroforestry the provision of multiple environmental services could be enhanced on agricultural lands (Boody et al. 2005; Swinton et al. 2006). Agroforestry systems, defined as the integration of trees, crops and animals, are sometimes viewed as a multi-functional working landscape that provide ecosystem services, environmental benefits and economic commodities (Jose 2009).

Agroforestry has had its share of sceptics in the past, though scientific data to substantiate its potential to generate various ecosystem services such as water quality improvement, biodiversity and soil fertility enhancement (amongst others, see Garrity 2004; Nair and Graetz 2004; Nair et al. 2009) have been steadily increasing. With growing environmental consciousness, agroforestry has gained increased attention as a potential sustainable alternative to traditional farming. The Millennium Ecosystem Assessment (2005) also emphasized the multi-functional role that agro-ecosystems can play in the provision of ecosystem services.

2.2 Pre-field Preparation

Preparation for the fieldwork was undertaken over the course of eleven months by the project members, composed of four doctoral and one master student who were supervised by five professors from various fields of expertise that included urban planning, landscape planning, urban engineering and public policy. Preparatory activities included a monthly progress evaluation sessions and Skype meetings with the local counterparts. The unit benefited much from collaboration with local partners, who provided assistance and access to various organizations and key informants, without which holding the unit would have been otherwise nearly impossible.

From the literature review to the field data collection, the study was grounded on the problem-based and solution-oriented framework advocated by sustainability science. In identifying the research problem, the initial literature review focused on the general critiques of PES implementation in Costa Rica. The findings of this review were then shared with local counterparts for feedback and to seek assistance in identifying the appropriate study site for an in-depth investigation.

This iterative process of formulating the research questions and objectives of the study was adopted to aptly reconcile the actual field situation with the gaps identified from literature. It was a time consuming process, but highly necessary for identifying the most relevant problem that could be addressed by the study given the limited time for the field work. The empirical knowledge gained by the authors during the actual field survey dynamically influenced the issues that the participants looked at, thus some of the issues identified during the initial literature review were excluded in the end.

The methodological approach, based on an iterative process of formulating the research questions and objectives of the study, was undertaken to ensure a match between knowledge gaps from literature with locally relevant issues. The adoption of this methodology was grounded in two reasons. First, although the Costa Rican PES is generally well studied, providing a rich resource for literature review, most studies were generated from data gathered between 1999 and 2005, generating a knowledge vacuum between the findings from the literature review and the current situation in Costa Rica. The second reason was the difficulty in establishing a new research project in an unfamiliar location. Since this was the first GFE Costa Rica unit to be implemented, narrowing down the topic and identifying the appropriate study site proved to be difficult.

The preliminary literature review revealed that the main criticisms on the Costa Rican PES has been the lack of additionality, or simply the argument that forest conservation in the country would have happened anyway without PES payments.

The initial research problem was therefore centered on how this lack of additionality could be addressed. The proposal was presented at the GPSS-GLI International Symposium in January 19, 2015, where the team received comments from the audience and the former Minister for the Environment and Energy of Costa Rica (1994–1998 and 2011–2014), the main proponent of PES during his first term.

On a follow-up interview on January 20, 2015, the former minister acknowledged some of the criticisms of PES, but at the same time disclosed that Costa Rican policy makers and PES implementers have already been devising ways to address them. As a result, a significant modification introduced recently was the implementation of PES for agroforestry systems. This approach potentially increases the additionality of PES, since it is implemented in agricultural lands, which presumably have a higher tendency for deforestation.

Building on this new finding, a targeted literature reviews on PES for agroforestry was conducted to develop the research questions for the field level survey. Since the policy goal for the program is to increase biodiversity and enable small holders to have access to PES, the survey questions focused on examining whether these goals have been achieved, as well as to identify difficulties faced by facilitators and implementers. The Research structure is summarized in Fig. 1.

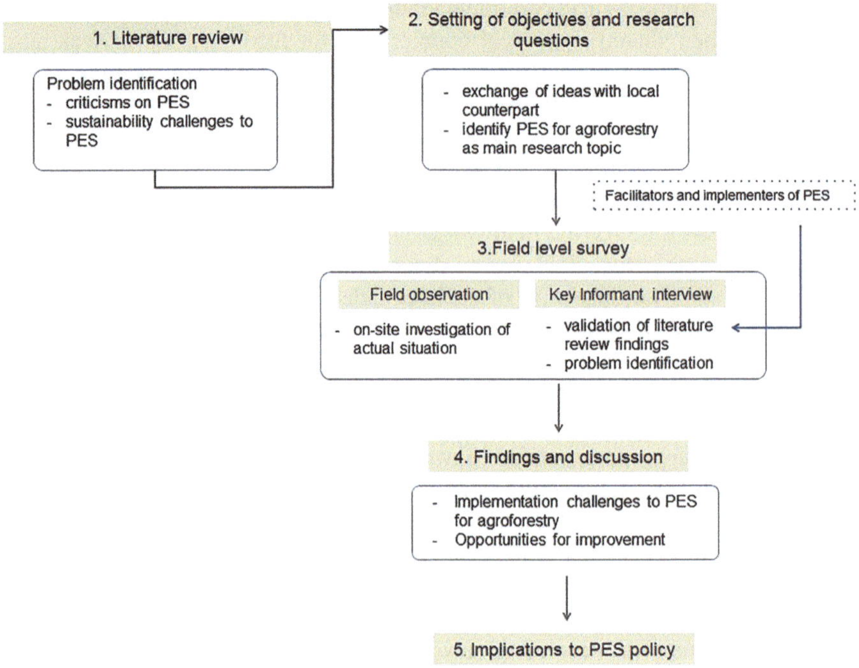

Fig. 1 Research structure

2.3 Fieldwork

Fieldwork was conducted from March 1 to 11, 2015 in San Jose City and Guanacaste Province (Fig. 2). San Jose is the capital city of Costa Rica and is the location of most offices visited to conduct stakeholder interviews. On the other hand, Guanacaste region was chosen as a study-site for field observations due to the high number of agroforestry projects in the region and the access to key informants from the Office of Guanacaste Conservation Area (ACG)—National System of Conservation Areas (SINAC). ACG is one of the eleven conservation areas that make up SINAC and is the only place in the world that contains four of the five major tropical systems (ACG-SINAC 2015).

Guanacaste is considered to be Costa Rica's most outstanding story of forest restoration, which provides a unique opportunity to study how and why the regrowth of tropical dry forest occurs (Calvo-Alvarado et al. 2009). In the past, Guanacaste's forest cover was almost eliminated, but is now currently being successfully restored. However, future pressures on Guanacaste's forest are being anticipated due to increased investment on tourism infrastructure and real estate development. Also regional expansion and intensification of cattle ranching, driven by rising beef prices in the international market, have significantly increased economic incentives to deforest. These threats highlight the need to integrate forest protection and incentives for sustainable agriculture through agroforestry for the sustainable development of the region.

The collection of field data was conducted primarily by qualitative methods that consisted of key informant interviews (formal and informal) and field observations. The interviewees and the institutions they represent were selected for the roles and functions they play in the implementation of PES (Table 1).

Fig. 2 Location of Costa Rica and GIS tracking of field survey route. *Source* Google earth administrative boundary (2015)

Table 1 List of stakeholders and organizational role in the implementation of PES for agroforestry

Stakeholder	Acronym	Sector	Roles/functions
National Forest Financing Fund 1. National office 2. Regional office	FONAFIFO	Government	• National—secures and administers PES funds • Regional—monitor and endorses PES contract
Guanacaste Conservation Area (ACG)—National System of Conservation Areas (SINAC)	ACG-SINAC	Government	• Legally responsible for control, monitoring and promotion of PES program
Tropical Agricultural Research and Higher Education Center	CATIE	Academia	• Provide technical assistance
International Union for Conservation of Nature	IUCN	NGO	• Provide technical assistance • Assist in securing funds
National Biodiversity Institute	InBio	NGO	• Provide technical assistance and expert advice on biodiversity issues
Coffee farmers' cooperative	COOPEDOTA	Private	• Agroforestry practitioners

3 Brief History on the Evolution of Costa Rican PES

To protect its remaining forested areas Costa Rica implemented a series of policies and legal actions for forest conservation and management. In 1969, the First Forest Law in the country regulated forest use on public land and established a system of national parks (Calvo-Alvarado et al. 2009). This extensive network of protected areas was intended to limit agricultural expansion into the forests (Fagan et al. 2013).

Also, there had been attempts to engage private landholders in forest recovery by providing incentives for reforestation and later for natural forests that included income tax reductions, subsidized credits, and transferable bonds (Brockett and Gottfried 2002). The reforestation subsidies were successful in reforesting 51,000 ha up to 1993, but were critiqued for benefitting mostly large holders, when it is the small holders who needed the subsidy for timber (Castro-Salazar and Arias-Murillo 1998).

In 1995, the reforestation incentives were eliminated as a result of the Structural Adjustment Program (SAPs) of the World Bank, except for one program targeting small holders. This led the forestry department to lobby for the revision of the Forestry Law in search of other funding sources, which also coincided with the post-Rio environmental movement. With strong support from the Ministry of Environment (MINAE) and the president of Costa Rica a Revised Forestry Law was passed, which launched the PES program in 1996 (Porras and Neves 2006).

The legal framework within which Cost Rica established its PES program rests on three laws (Sánchez-Azofeifa et al. 2007). First is the Environment Law 7554 of 1995 that mandates a "*balanced and ecologically driven environment*" for all. Second is the 1996 Revised Forestry Law 7575 prohibiting land cover change in forests, mandating a "rational use" of all natural resources. Finally, the Biodiversity Law, enacted in 1998, promotes the conservation and "rational use" of biodiversity resources. The passage of these laws reflects the change in the mindset of Costa Rican policy makers, resulting from increased recognition of the need to put a stop to the environmental degradation that threatens the sustainable supply of services provided by forests.

The Forestry Law 7575 also formalized the National Forest Financing Fund (FONAFIFO), an institution entrusted with managing and acquiring funds for PES. The funds administered by FONAFIFO come from various sources, both domestic and foreign. Private funds at the beginning of the program were mostly sourced from sales of carbon bonds to national and international organizations e.g. carbon certificates sold to Norway in 1997. Domestic funding comes mainly from the one third of the sales tax on fossil fuel, as mandated by the same forestry law.

To expand FONAFIFO's budget, a project called *Ecomarkets I* (amounting to US$ 49 million, 65 % of which was loaned from the World Bank) was used to fund 50 % of PES contracts from 1996 to 2005 (Sills et al. 2005). This was complemented by donations from local hydropower companies and the German Development Bank. During the second phase of the project (*Ecomarkets II*), which took place from 2007 to 2011, the Costa Rican government bore 52 % of the total funding, a significantly larger portion than in *Ecomarkets I*.

The nature of the Costa Rican PES funding, which in its early stages comes mostly from grants and donations, has resulted in some criticism in the past. Some articles referred to it as a subsidy program (Bennett 2010) and PES-like scheme (Wunder 2008a, b) due to the reason that the national government acts as buyer in behalf of environmental users. However, the recent introduction of private purchasing agreements for watershed/forest protection and carbon footprint offsetting so that FONAFIFO can receive funding from institutions and companies benefitting from PES has improved the image of the Costa Rican PES. A national bill in 2006, which sets a water tariff on those who use water as a "raw material" in their production activities to fund watershed protection projects (Murillo et al. 2012), reinforces the concept of PES in the country as a mechanism for internalizing negative environmental externalities by compensating those who ensure the provision of environmental services.

When an individual is accepted for participation in the PES program, a contract is established between the landholder and FONAFIFO. In return, the landholders cede their carbon and other environmental rights to FONAFIFO for the length of the contract, but are afterwards free to renegotiate the prices, or sell the rights to other parties (Karousakis 2007). The first payment is granted during the signing of contract, though succeeding payments are disbursed on an annual basis only after verification that they are following their contracts (Pagiola 2006). Monitoring is essential to ensure that the program attains its intended effect and that PES payments are indeed promoting desirable land use changes.

The Costa Rican PES compensates landholders who maintain either planted or protected forests in their land in order to ensure the continuous provision of ecosystem services. It is a market-oriented approach that aims to compensate for the positive externalities (environmental services) provided by forests. PES is in principle an effective way to induce conservation while compensating those who incur its costs (Pfaff et al. 2008). It is hoped that this additional income in the form of PES payments will transform the long-standing debate as to whether sustainable forestry can be economically competitive with alternatives such as conversion to pasture or plantations (Kishor and Constantino 1993). The argument is that if forest owners get compensated, the additional income might be enough to make private forest conservation economically viable and thus attractive to local landowners.

For years, the Costa Rican PES has been regarded as one of, if not the most successful conservation story, which attracted the attention of many scholars, who have revealed contrasting views and opinions on its perceived impact. Besides impact evaluation, the research interest on the potential linkages between PES and poverty has also grown. In recent years, policy makers and researchers have started discussions on how PES programs can be designed to maximize poverty reduction. Of particular interest is the identification of trade-offs between efficiently generating environmental services and reducing poverty (Pagiola et al. 2005). While PES programs are not specifically designed to be a poverty reduction program, they have the opportunity to do so given the high spatial correlation between economically disadvantaged areas and areas that supply environmental services (Wunder 2008a, b; Pagiola et al. 2005).

To address criticisms that have been discussed earlier, some modifications on the design of PES were introduced over the course of its implementation, which includes the addition of agroforestry systems and "assisted" natural regeneration to the existing land use modalities. See also Table 2 for a complete list of PES modalities. At the beginning, the program promoted three modalities: forest protection, reforestation, and sustainable forest management. Forest protection requires forest owners to protect existing primary or secondary forests for 5 years. Reforestation contracts entail the planting of trees on agricultural or abandoned land, which has to be maintained for 15 years. Sustainable forest management compensated landowners with approved a "sustainable logging plan". This program aims to conduct low-intensity logging while keeping forest service's intact.

4 Payments for Agroforestry Systems: A Tool to Increase Supply of Environmental Services from Agricultural Lands

The payment for agroforestry contracts was introduced in 2003 to incentivize the planting of trees on farmlands. Activities may include afforestation of pasturelands or shade coffee and live hedges providing fodder or windbreaks (Porras and Neves 2006). It aims to boost biodiversity in agricultural lands as well as to increase the reach of PES to farmers with smallholdings (FONAFIFO 2013).

Table 2 List of PES modalities by year of implementation from 1997 to 2015 (FONAFIFO 2015)

MODALITIES	1997	1998	1999	2000	2001	2002	2003	2004	2005	2006	2007	2008	2009	2010	2011	2012	2013	2014	2015
Established plantations		■	■				■	■											
Forest management	■	■	■		■	■								■	■	■	■	■	■
Forest protection	■	■	■	■	■	■	■	■	■	■	■	■	■	■	■	■	■	■	■
Reforestation	■	■	■	■	■	■	■	■	■	■	■	■	■	■	■	■	■	■	■
Agroforestry systems							■	■	■	■	■	■	■	■	■	■	■	■	■
Reforestation in second crops								■	■	■	■	■	■	■	■	■	■	■	■
Protection of water resource								■	■	■	■	■	■	■	■	■	■	■	■
Reforestation with endangered species										■	■	■	■	■	■	■	■	■	■
Reforestation (multi-stage)										■									
Natural regeneration (productive)										■	■	■	■	■	■	■	■	■	■
Natural regeneration projects (MDL)										■	■	■	■	■	■	■			
Natural regeneration of pasturelands													■	■	■	■	■	■	■
Forest protection in disaster areas													■						
Coffee agroforestry systems													■	■	■	■	■	■	■
Protection of wild areas													■	■	■	■	■	■	■
Protection of conservation gaps													■	■	■	■	■	■	■
Agroforestry systems with endangered species															■	■	■	■	■
Agroforestry systems with native species															■	■			
Trees in coffee plantation																	■	■	■

Integrating biodiversity goals with sustainable agriculture provides an opportunity to justify funding for biodiversity. Because there is no identifiable single beneficiary of biodiversity, but rather the service is a public and universal good, funds for biodiversity have historically come from one-time grants (Bennett and Henninger 2009), making them difficult to sustain in the long-term.

4.1 Payment Scheme and Basic Requirement

Payment for agroforestry systems in an attempt to make it more accessible to smallholders and is the only modality that is payable by the number of trees, significantly different to other PES modalities which are paid by hectare of land (see also Table 3).

The basic contract for agroforestry pays for 350 trees up to a maximum of 3,500 trees per landowner, though indigenous people that share land ownership rights are allowed a maximum of 1,000 ha/year or 30,000 trees per community. Local non-governmental organizations that function as intermediaries between smallholders and authorities are not subjected to an area limit. Besides

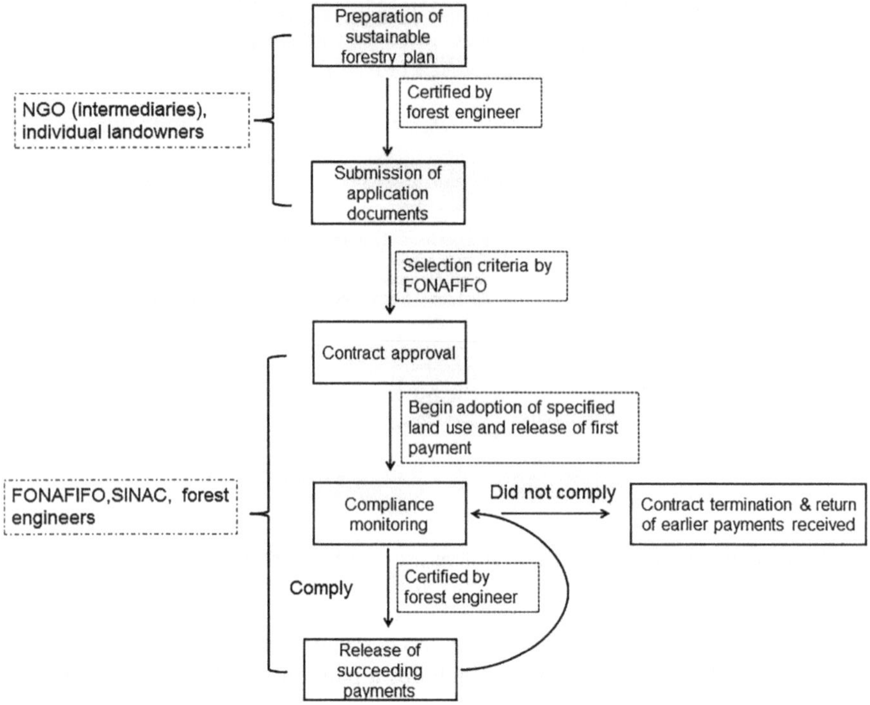

Fig. 4 Stages of implementation of PES for agroforestry. By authors based on interview data

is prepared by the forest engineer. With the aid of a state of the art database and through the conduct of a regular audit, FONAFIFO is capable of tracking the compliance of participants as well as whether the certifications by forest engineers are correct (Pagiola 2006). Non-compliance of participants will result in the forfeiture of future payments and the responsibility to return previous payments. Forest engineers on the other hand can also lose their license over a false certification of compliance. The stages of implementation of PES for agroforestry are also summarized in Fig. 4.

5 Findings and Discussion on Emerging Issues in Implementation

5.1 Challenges to Implementation: Focus on Quantity Over Quality

The results of the study suggest that PES for agroforestry is currently focused on increasing forest cover rather than ensuring the quality of ecosystem services generated. During the interview with ACG-SINAC (March 9, 2015), the informants

Fig. 5 Interview at FONAFIFO national headquarters, San Jose City (March 4, 2015)

criticized the current PES scheme as being solely focused on increasing the number of standing trees. FONAFIFO officials (see Fig. 5, March 4, 2015) acknowledged that the primary intention of the PES for agroforestry is to achieve the remaining portion of the country's 2020 forest cover target. In order to do so, the government designed PES for agroforestry systems so that it can address the shortcomings of the previous modalities, which limited participants to those having at least one hectare of land (PES for agroforestry provides payment not by land area but by number of trees planted).

However, while this payment scheme enabled the program to increase its reach in areas not previously accessed by first-generation land use modalities, it also created perverse incentive for landowners. CATIE agroforestry experts (see Fig. 6, March 2, 2015) noted that some farmers have little interest in maintaining a good quality agroforestry farm, and are only motivated to increase the number of trees planted so that they receive larger payments. Also, they pointed out that planting more trees does not necessarily translate to an increase in biodiversity. Thus, it can be argued, the present policy has compromised the quality of the agroforestry systems, which in turn impacts the optimization of environmental services. A similar finding by Pagiola (2006) also suggests that paying for greater land cover does not necessarily increase the flow of ecosystem services.

Furthermore, CATIE representatives (March 2, 2015) claimed that local farmers have inadequate knowledge of good agroforestry practices, such as appropriate planting density and tree combinations, which also contributes to the poor quality of agroforestry farms. Similar observations were echoed during the field visit

Fig. 6 Interview with CATIE agroforestry experts. Photo taken in CATIE campus, Turrialba, Costa Rica, March 2, 2015

to Guanacaste. The FONAFIFO—regional office representative (see Fig. 7, March 2015) pointed out that the difficulty in monitoring compliance and providing technical assistance to farmers due to a shortage of field staff exacerbates this problem.

On the other hand, it can also be surmised that the inadequate knowledge of farmers in agroforestry requirements can be attributed to a lack of access to necessary information and training. Based on the current implementation design of PES for agroforestry, local landowners may enter into a contract without direct contact with FONAFIFO. Intermediary agencies—through forest engineers—can prepare the forestry plan for farmers and proceed with applications on behalf of landowners. Given the vast number of farmers that these agencies deal with, there is a high possibility that farmers do not get all the information they need. One possible scenario could be that they are unaware of what is expected from them (as technical specifications for planting distance or recommended tree combination might not actually get distributed to them).

Another challenge according to the FONAFIFO regional officer (March 7, 2015) is that some agroforestry farms cannot be strictly monitored, as they are located in remote areas that are difficult to access even by car. During ocular inspections of agroforestry farms, inappropriate planting density, incompatibility between crops and trees planted, and improper practices such as barbed wires cutting into the tree fence were documented by the authors (see Figs. 8 and 9). According to CATIE experts (in an interview on March 2, 2015) these improper practices affect the quality of timber produced and do not contribute to the development of integrated and sustainable agroforestry system.

Fig. 7 Field visit to an agroforestry site in Guanacaste. Taken in March 7, 2015

Fig. 8 Agroforestry plot in Guanacaste showing improper planting distance. Photo taken on March 7, 2015

Fig. 9 Live hedges with barbed wires cutting into the trees. Photo taken on March 7, 2015

Another factor that contributes to farmers' half-hearted commitment to agroforestry were differences between the ministry of agriculture and forestry (as discussed with a ACG-SINAC staff officer on March 9, 2015). According to such explanation these different ministries pursue different goals, which sometimes can sometimes create confusion amongst farmers. In addition, the structural separation of agriculture and forestry hinders the implementation of joint technical assistance programs and activities. A CATIE representative (March 4, 2015) made the same observation and noted that Costa Rica still has no law on agroforestry synergy between forestry and agriculture, which often leads to organizational conflict. The CATIE officer gave an example where those involved in shade coffee production complain that the recommended tree species cannot protect coffee trees from strong rain. Other farmers also complain about the felling of trees necessary to maintain a good agroforestry system, which they perceive destroys the beauty of their farm as well as causing damage to crops. On the other hand, cattle farmers are concerned about the appropriate ratio of cattle to the number of trees that they can grow.

The lack of sufficient information can be a strong deterrent for participation in PES, as well as an impediment to the development of a multifunctional agroforestry system. The manager of Coope Dota, a farmer's cooperative (March 2, 2015), revealed to the authors that only a handful of their members are enrolled in the program, despite the fact that most of their members have been traditionally practicing agroforestry prior to the implementation of PES. The main reason for this appears to be gaps between the technical specifications required by PES

Fig. 10 Coope Dota agroforestry experimental plot. Photo taken March 2, 2015

for agroforestry and the traditional agroforestry knowledge of farmers (related for example to the planting distance). In this sense, the cooperative is currently doing its own research on the ideal coffee shade combination through an experimental farm (see Fig. 10).

5.2 Opportunities for Improvement

Based on the findings of the present research there appear to be various opportunities to improve the quality of agroforestry systems, which in turn would ensure the sustainable generation of ecosystem services. One of the most direct ways is to enhance the technical knowledge of farmers on agroforestry. CATIE officials (March 4, 2015) suggested that in order to improve implementation, technical assistance to farmers should be made available and accessible in order to educate them on good practices and benefits of a fully functioning agroforestry farm.

One opportunity to do so can be during the preparation of the forestry plan, when the landowner directly interacts with the forest engineer. For example, prior to the preparation of the actual plan, a pre-qualifying stage where forest engineers assess the capabilities of local farmers in managing agroforestry farms could be added. This pre-qualification process could also include a baseline evaluation of applicants' knowledge and skills in establishing and maintaining an agroforestry system.

In addition, the auditing procedure could be improved by increasing the number of monitoring visits. The auditing team is composed of representatives from SINAC, FONAFIFO and forest engineers. According to the FONAFIFO regional staff (March 7, 2015) the team conducts scheduled visits to agroforestry farms to verify annual compliance reports by forest engineers. These compliance reports are prepared and submitted to FONAFIFO by forest engineers. The payments for PES are staggered over a span of 3 years and are only released once compliance reports are received and certified by FONAFIFO. Making unannounced field visits, aside from regular auditing visits, or increasing the number of scheduled monitoring could help improve the diligence of agroforestry farmers.

On the other hand, to address the perverse incentive arising from payment being made according to the number of trees planted, a performance-based payment scheme should be designed as soon as possible to deter farmers from focusing only on the quantity of trees planted but rather on actual provision of environmental services. Although designing PES programs based on performance is as much a socioeconomic challenge as it is a scientific one (Bohlen et al. 2009), the active engagement of stakeholders with technical expertise could provide a valuable opportunity to overcome such challenges. For example, InBio and IUCN, another international NGO working closely with Costa Rica can assist in creating practical guidelines for the valuation of biodiversity services and greenhouse gas mitigation, respectively. In this way, farmers whose agroforestry plots generate higher rates of ecosystem services could get more compensation. This incentive system can help ensure the development of a multi-functional agroforestry system.

The greatest opportunity for the improvement of PES for agroforestry lies in the inclusive design of PES itself. While the Costa Rican PES is a state-driven scheme, different groups can take the lead in developing PES through different activities (Bennett and Henninger 2009). Some of the institutions interviewed for this study do not have a legal mandate to engage in PES (for example CATIE, IUCN and InBio), yet they continue to play an active role in providing technical advice for the improvement of the system.

However, in order to sustain such collaboration, providing incentives for these external institutions is clearly necessary. Conducting joint projects between FONAFIFO and these institutions could be a way to incentivize participation. For example, CATIE, having expertise on agroforestry, could be tapped to create agroforestry modules that will be disseminated to intermediary agencies or even directly to landowners. The realization of a multi-functional agroforestry system that is not solely dependent on PES payments to be sustainable can easily be achieved if the resources of all-important stakeholders are combined to achieve a shared vision.

Besides the inclusive design of PES, the flexible nature of the law itself provides an opportunity for the program to be able to adapt to the changing needs and conditions of landowners. According to FONAFIFO (2013), for this exact purpose the law on PES was prevented from being overly prescriptive, so that implementing regulations can be changed when needed. This system allows for the program to cope with changing political climate as well as fund availability. The presidential term in Costa Rica lasts only for four years and before the year 2005 an elected president could not run for a second term. Based on this premise, PES proponents in 1997 designed it to be flexible as a safeguard to ensure the continuity of PES despite predictable change in government priorities as each new president was sworn into office. In this sense, the executive decree issued annually by MINAE identifies priority criteria for each modality of PES, including how many hectares are eligible for funding and how much is to be paid for each modality, making the program flexible to changing needs. This flexibility allows for the program to implement changes within a short period of time, ensuring an efficient use of funds. For example, during the initial implementation of PES for agroforestry the amount of payment for each tree was constant and did differentiate between native and foreign tree species. However, in 2014 FONAFIFO started to implement a higher payment for native species in danger of extinction.

6 Implications for the Future Development of PES

The willingness and capacity of the Costa Rican government (through FONAFIFO) to modify existing rules and regulations to match PES policy with the changing needs of landowners supports earlier studies' claim that the success of the PES rests in its adaptable design. Having a flexible law and flexible people, both government and nongovernment officials, in place to make changes when needed has proven instrumental to Costa Rica's success (Bennett and Henninger 2009). The adaptable nature of Costa Rican PES can be attributed to the fact that it was designed with a long-term vision. With the awareness that political support is not permanent, the proponents of PES crafted policies and institutions that transcended presidential terms. Through the annual issuance of executive decrees for the implementation of PES, modifications to improve it (such as introduction of new modalities and changes in prioritization criteria) can be implemented within a short period of time.

The addition of agroforestry as one of the modalities of PES was a timely and significant modification in the design of the program to address its criticisms. However, findings of this study show that the implementation of PES for agroforestry can still be improved, in order to satisfy its goal of ensuring the sustainable provision of environmental services. While PES for agroforestry fulfils the short-term goal of increasing the number of trees, focusing on the number of trees alone without regard to the quality of the system does not guarantee the generation of sustainable ecosystem services. Experts from CATIE coined the term "political forests" in reference to planting trees for the sake of receiving cash incentives, without regard to the quality of resulting agroforestry farms. This situation compromises the long-term sustainability of the program, as participation becomes solely dependent on the availability of cash incentives.

To improve implementation, a performance-based payment scheme could improve the current payment scheme (based on number of trees planted) to deter opportunistic behaviour by landowners. However, designing a PES program based on performance is challenging both in scientific and economic terms. The conduct of a baseline assessment of the current level of environmental services being provided and future potentials needs both technical and financial resources. In the case of Costa Rica, the technical expertise of CATIE, IUCN and InBio can be tapped to develop a cost-effective documentation scheme for environmental services generated from the program. This would enable the environmental service sellers to know the actual value of ecosystem services they should be compensated for, and the buyers to secure proof that they are getting their money's worth. The success of this scheme could lead to innovations on cost-efficient ways for ecosystem service provision.

Stricter guidelines for forestry engineers in the preparation of sustainable forestry plan as well as monitoring compliance could also further improve the system. Furthermore, streamlining technical assistance to landowners from various stakeholders including academia, non-governmental organizations (NGOs) and cooperatives is recommended to improve their knowledge and understanding of the underlying ecology needed for the establishment and maintenance of a multifunctional agroforestry system.

Finally, the authors would like to note that one of the limitations of the present chapter lies in the fact that the assessment of the program was only carried out from the perspective of the demand side, whilst that the supply side was not incorporated in the discussion. Although CATIE, ACG-SINAC and FONAFIFO employees provided useful landowner insights -based on years of field experience- further research is recommended in this direction.

References

ACG-SINAC. (2015). Area de Conservacion Guanacaste-Sistema Nacional de Areas de Conservacion. Retrieved October 29, 2015, from http://www.acguanacaste.ac.cr/acg/que-es-el-acg.

Batie, S. (2008). Wicked problems and applied economics. *American Journal of Agricultural Economics, 90*(5), 1176–91. Retrieved November 11, 2015, from http://ajae.oxfordjournals.org/cgi/doi/10.1111/j.1467-8276.2008.01202.x.

Bennett, K., & Henninger, N. (2009). Payments for ecosystem services in Costa Rica and Forest Law No. 7575: Key Lessons for Legislators. Retrieved October 28, 2015, from http://www.agora-parl.org/sites/default/files/090422_e-parliament_forests_initiative_0.pdf.

Bennett, K. (2010). Additionality: The next step for ecosystem service markets. *Duke Environmental Law and Policy Forum, 20*(2), 417–439.

Bohlen, P. J., Lynch, S., Shabman, L., Clark, M., Shukla, S., & Swain, H. (2009). Paying for environmental services from agricultural lands: an example from the northern Everglades. *Frontiers in Ecology and the Environment, 7*(1), 46–55.

Boody, T. C., Vondracek, B., & Andow, D. (2005). Multifunctional agriculture in the US. *BioScience, 55*, 27–48.

Brockett, C. D., & Gottfried, R. R. (2002). State policies and the preservation of forest cover: Lessons from contrasting public-policy regimes in Costa Rica. *Latin American Research Review, 37*, 7–40.

Calvo-Alvarado, J., McLennan, B., Sánchez-Azofeifa, A., & Garvin, T. (2009). Deforestation and forest restoration in Guanacaste, Costa Rica: Putting conservation policies in context. *Forest Ecology and Management, 258*, 931–940. Retrived September 15, 2015, from http://doi.org/10.1016/j.foreco.2008.10.035.

Castro-Salazar, R., & Arias-Murillo, G. (1998). *Costa Rica: Toward the sustainability of its forest resources (Technical report)*. San Jose, Costa Rica: Fondo Nacional de Financiamiento Forestal.

Clark, W. (2006). Sustainability science: A room of its own. *Proceedings of the National Academy of Sciences in the United States of America (PNAS), 104*, 1737–1738.

Clay, J. (2004). *World agriculture and the environment: A commodity -by-commodity guide to impacts and practices*. Washington, DC: Island Press.

Fagan, M. E., DeFries, R. S., Sesnie, S. E., Arroyo, J. P., Walker, W., & Soto, C., et al. (2013). Land cover dynamics following a deforestation ban in northern Costa Rica. *Environmental Research Letters, 8*, 034017. Retrieved September 15, 2015, from http://doi.org/10.1088/1748-9326/8/3/034017.

FONAFIFO. (2013). The fund for sustainable biodiversity: Investing today for a better future. In A. Vilma Obando, G. Jesus Ugalde, A. Herrera V (Eds.) (1st ed.).

FONAFIFO. (2015). Environmental Services Payment (PSA) Program. Retrieved August 11, 2015, from http://www.fonafifo.go.cr/home/psa_eng/index.html.

Garrity, D. P. (2004). Agroforestry and the achievement of the Millennium Development Goals. *Agroforestry Systems, 61*(1–3), 5–17.

Horn, R. E., & Weber, R. P. (2007). *New tools for resolving wicked problems: Mess mapping and resolution mapping processes*. Watertown, MA: Strategy Kinetics L.L.C.

Jose, Shibu. (2009). Agroforestry for ecosystem services and environmental benefits: An overview. *Agroforestry Systems, 76*(1), 1–10.

Karousakis, K. (2007). Incentives to reduce GHG emissions from deforestation: Lessons learned from Costa Rica and Mexico. Retrieved September 15, 2015, from http://www.oecd.org/dataoecd/55/54/38523758.pdf\nwww.oecd.org/env/cc/aixg.

Kishor, N., & Constantino, L. (1993). Forest management and competing land uses: An economic analysis for Costa Rica. Laten Dissemination Note no. 7. Washington, D.C.: World Bank. Retrieved September 14, 2015, from http://www-wds.worldbank.org/external/default/WDSContentServer/WDSP/IB/2000/01/06/000094946_99122006050427/Rendered/PDF/multi_page.pdf.

Millennium Ecosystem Assessment. (2005). *Ecosystems and human well-being: Biodiversity synthesis*. Washington, D.C.: World Resources Institute.

Murillo, R., Kilian, B., & Castro, R. (2012). *Leveraging and sustainability of PES* (p. 267). Ecosystem Services from Agriculture and Agroforestry: Measurement and Payment.

Nair, V. D., & Graetz, D. A. (2004). Agroforestry as an approach to minimizing nutrient loss from heavily fertilized soils: The Florida experience. *Agroforestry Systems, 61*(1–3), 269–279.

Nair, P. R., Kumar, B. M., & Nair, V. D. (2009). Agroforestry as a strategy for carbon sequestration. *Journal of Plant Nutrition and Soil Science, 172*(1), 10–23.

Pagiola, S., Arcenas, A., & Platais, G. (2005). Can payments for environmental services help reduce poverty? An exploration of the issues and the evidence to date from Latin America. *World Development, 33*(2), 237–253. Retrieved Sepetmber 16, 2015 from http://doi.org/10.1016/j.worlddev.2004.07.011.

Pagiola, S. (2006). Payments for environmental services in Costa Rica. Retrieved August 30, 2015, from https://mpra.ub.uni-muenchen.de/2010/1/MPRA_paper_2010.pdf.

Pfaff, A., Robalino, J. A., & Sanchez-Azofeifa, G. A. (2008). *Payments for environmental services: empirical analysis for Costa Rica*. Duke University, Durham, NC, USA: Terry Sanford Institute of Public Policy.

Porras, I., & Neves, N. (2006). Costa Rica National PES Programme. In *Case Study. London: Watershed Markets*. Retrieved September 14, 2015, from http://www.watershedmarkets.org/casestudies/Costa_Rica_National_PES_eng.html.

Rittel, H., & Webber, M. (1973). Dilemmas in a general theory of planning. *Elsevier Policy Sciences, 4*, 155–169.

Sánchez-Azofeifa, G. A., Pfaff, A., Robalino, J. A., & Boomhower, J. P. (2007). Costa Rica's payment for environmental services program: intention, implementation, and impact. *Conservation Biology, 21*(5), 1165–1173.

Sills, E., Hartshorn, G., Ferraro, P., & Spergel, B. (2005). *Evaluation of the World Bank-GEF Ecomarkets Project in Costa Rica.* Panel Evaluation Report, North Carolina University, United states and the World Bank, Washington, DC. Retrieved October 29, 2015, from http://www2.gsu.edu/~wwwcec/docs/doc%20updates/NCSU_Blue_Ribbon_Panel_Final.pdf.

Swinton, M., Lupi, F., Robertson, G. P., & Landis, D. A. (2006). Ecosystem services from agriculture: Looking beyond the usual suspects. *American Journal of Agricultural Economics, 88*, 1160–1166.

Wunder, S. (2005). Payments for environmental services: Some nuts and bolts. *CIFOR Occasional Paper, 42*(24), 1–14.

Wunder, S. (2008a). Payments for environmental services and the poor: Concepts and preliminary evidence. *Environment and Development Economics, 13*:279–297.

Wunder, S. (2008b). Necessary conditions for ecosystem service. In *Economics and conservation in the tropics : A strategic dialogue* (pp. 1–10).

Part III
Exercises on Resilience

Assessment of Fieldwork Methodologies for Educational Purposes in Sustainability Science: Exercise on Resilience, Tohoku Unit 2015

Ricardo O. San Carlos, Olga Tyunina, Yuki Yoshida, Aimee Mori, Giles Bruno Sioen and Jiaqi Yang

Abstract This chapter introduces the Tohoku Unit, a GPSS-GLI Exercise on Resilience conducted in the northeastern area of the Honshu Island in Japan. This unit consists of a one-week educational fieldwork, where students are exposed to the reconstruction process in Otsuchi Town, Iwate Prefecture that has followed the Tohoku Earthquake Tsunami of March 2011. In the present chapter an assessment of the field exercise is presented in two levels of depth, (1) an application of field methodologies to identify sustainability issues in the reconstruction process, and (2) the contribution of the fieldwork to the development of student's competencies relevant to sustainability research, following the framework of key competencies in sustainability proposed by Wiek et al. (Sustainability Science 6(2):203–218, 2011a; Environment 53(2):3–13, 2011b). The authors discuss the relationship between the field methodologies utilized and the results obtained in the two levels of the assessment. The results indicate that the field methodologies utilized contributed to a high number of reconstruction issues being identified, with the majority of issues belonging to the social dimension. Also, the different lectures attended during the field work and preparatory work highlighted the highest number of issues identified amongst all three sustainability dimensions (social, economic, environmental). Regarding the competencies, students seem to recognize the importance of the five competencies suggested by Wiek et al. (Sustainability Science 6(2):203–218, 2011a; Environment 53(2):3–13, 2011b), and overall consider the contribution of the field exercise as "satisfactory" to "effective" in

R.O. San Carlos (✉) · O. Tyunina · Y. Yoshida · A. Mori · G.B. Sioen · J. Yang
Graduate Program in Sustainability Science-Global Leadership Initiative,
Graduate School of Frontier Sciences, The University of Tokyo,
5-1-5 Kashiwanoha, Kashiwa, Chiba 277-8563, Japan
e-mail: sancarlos.ricardo@gmail.com

M. Esteban et al. (eds.), *Sustainability Science: Field Methods and Exercises*,
DOI 10.1007/978-3-319-32930-7_4

promoting these competencies. Final remarks highlight the relevance of considering this framework for future improvements in the design of the units by considering the main competencies that they intend to foster in the students. Also, the authors highlight some of the limitations of the framework in capturing the whole experience of students in the field, and the need to incorporate a deeper analysis of the concept of resilience in future assessments of this field exercise.

Keywords Sustainability Education · Assessment · Field Exercise · Competencies

1 Introduction

In 2012, the University of Tokyo founded the Graduate Program in Sustainability Science—Global Leadership Initiative (GPSS-GLI) as part of the Graduate School of Frontier Sciences (GSFS). The program strives to foster future leaders, who not only have a deep and specialized understanding of a specific field but also a broad knowledge base, to tackle the variety of social, environmental, and ecological issues that societies face today (GPSS-GLI 2015). GPSS-GLI has three main objectives in both its Masters and PhD programs: "to develop the skills necessary for global leadership, to acquire a broad perspective and problem solving capabilities, and to learn to apply the concept of "resilience" both theoretically and practically" (GPSS-GLI 2015). In order to achieve these objectives, the program emphasizes the importance of achieving a holistic understanding, looking to build a resilient society that absorbs shocks, and approaching issues in a trans-boundary manner. One of the characteristics of GPSS-GLI is that it offers practical education and real world training through field exercises. The Exercises on Resilience (ER) have been designed as a specific part of the GPSS-GLI curriculum, focusing on giving students a fieldwork experience that would contribute to the understanding of the concept, applied to real issues in sustainability. The Tohoku Unit is one of the ER and was established in 2013 after Japan was hit by the Tohoku Earthquake and Tsunami that took place on March 11th, 2011. The unit takes the recovery process of the Tohoku area as a case study, with the aim of showing students the role that resilience plays in the building of a sustainable community.

As few assessments have so far been conducted on the effectiveness of GPSS-GLI and other programs in developing education for sustainability, this chapter strives to make some headway by evaluating the field methodologies of a field exercise in GPSS-GLI. A direct indicator of students' knowledge and capabilities, competencies are meaningful outcome measures of education (Voorhees 2001). Accordingly, educators around the world are looking to transition from the traditional focus on assessment as the objective of learning to learning as the objective of education (e.g., Estrada 2014; Baartman et al. 2007). Moreover, it is well-established that the unique challenges in the field of sustainability require its practitioners to possess unique competencies (Willard et al. 2010; Rowe 2007), and education in sustainability is focusing increasingly on the development of such competencies (e.g., International Commission on Education for Sustainable

Development Practice 2008). Thus, the assessment of educational programs based on a common and explicit understanding of the necessary competencies not only have direct relevance to the labor market (Voorhees 2001), but are transparent and have the potential to facilitate the design of coherent and innovative academic programs.

Although sustainability science is still young as an academic field, some attempts have already been made to conceptualize the output of newly emerged educational programs. For example, De Haan (2006) introduces a list of eight sub-competencies critical for education for sustainable development when describing the German BLK '21' Programme, which worked on launching Education for Sustainable Development into schools (De Haan 2006). Segalas et al. (2009) suggest a grouping into knowledge and understanding competence, skills and abilities competence and attitudes competence, and discussing barriers in finding unified definitions of competencies that would enable rigorous comparison between several universities (Segalas et al. 2009).

Given the distinctive feature of how GPSS-GLI attempts to focus in training for real-world problem solving (GPSS-GLI 2015), the authors utilize the key competencies in sustainability described by Wiek et al. (2011a, b). This framework provides a concise list of five key competencies in the context of a research and problem-solving framework. According to Wiek et al. (2011a, b) these competencies are necessary to make sense of complex issues and systems, foresee possible sustainability visions, apply solutions strategically, have an understanding of the normative dimension of the issues, and to do so collaboratively. This problem-solving framework in which the competencies have been defined seems to be appropriate to address the nature of GPSS-GLI field exercises.

The rest of the chapter is structured as follows: The second section describes the overall structure and results of the authors' assessment of the field methodologies utilized in the unit, considering two levels of depth (1) the direct application of field research methodologies to identify issues in the reconstruction process, and (2) the educational impact of these field research methodologies on the development of key competencies amongst the participants of the field exercise. In the third section the results are discussed, relating the field research methodologies implemented with the issues identified (Level 1) and the competencies developed during the unit (Level 2). Finally, in the light of the assessment conducted by the authors, some concluding remarks and observations are presented.

2 Methodologies and Assessment of the Exercise on Resilience, Tohoku Unit 2015

The Exercises on Resilience have a predominantly educational aim, rather than being focused on research (and thus being slightly different to other field exercises in GPSS-GLI), such as the GFEs (see Chap. 1 in the present book). In the one-week during which they take place, these exercises concentrate on the understanding of the complexities of sustainability issues on the field. In this chapter, the authors have decided to focus on the application of these methodologies for educational purposes within a sustainability science graduate program. However, it is worth noting at this point that neither the unit nor the graduate program were designed with the objective of furthering the specific competencies discussed in this chapter, and readers should be cautioned against viewing the results of this analysis as an evaluation of the overall educational value of this exercise. Moreover, the intention and emphasis of this analysis was to conduct a reflective exploration rather than to test a predetermined hypothesis.

The authors identified two main levels of expected learning outcomes for the Tohoku Unit field exercise within the context of GPSS-GLI.

- Level 1: Outcome of fieldwork
  - This first level deals with the direct outcome from the application of research methodologies during the Tohoku Unit, specifically related to the identification of issues related to the reconstruction process in Otsuchi Town and surrounding areas.
- Level 2: Development of competencies for sustainability research
  - The second level relates to the contribution of the field exercise in the context of an educational sustainability program, expressed as its effectiveness in the development of key competencies for research in sustainability amongst its participants.

As the term 'sustainability' gains attention in academic circles, the discussion about sustainability programs' learning outcomes and key competencies for sustainability research and problem solving is unfolding in literature (Shephard 2008; Mochizuki and Fadeeva 2010; Frisk and Larson 2011; Wiek et al. 2011a). In a broader context, the expression 'sustainability literacy' is sometimes used (Ansari and Stibbe 2009). Proposed lists of key competencies for sustainability often vary in terms of content and lack consistency in naming. At the same time, overlaps between them clearly indicate that there are some common trends. For example, holistic or system thinking competence, long-term or foresighted thinking competence, collaboration or cosmopolitan perception competence, as well as interdisciplinary competence, are often listed as core competencies. Action-oriented skills (e.g., planning, implementation), value-oriented skills (awareness of different ethical values, cultural literacy) as well as psychological capacities (e.g., for empathy and compassion) are sometimes added (Barth et al. 2007).

Creating a comprehensive checklist of competencies for sustainability research and problem solving represents a challenging task and is beyond the scope of this work. For this reason, the framework introduced by Wiek et al. (2011a, b) was utilized as a basis for the discussion related to the Level 2 analysis carried out in the present work (Fig. 1). This suggests five key competencies, namely, system-thinking competence, anticipatory competence, normative competence, strategic competence and interpersonal competence that are highly relevant to the field of sustainability (for a description of the five competencies see Table 1).

Fig. 1 Two-level assessment of exercise on resilience

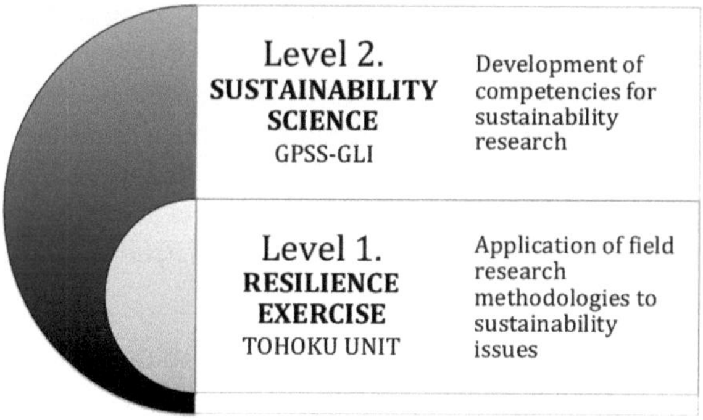

Table 1 Definition of competencies in sustainability (Wiek et al. 2011a) and their application to ER Tohoku 2015

Competencies	Definition (Wiek 2011)	Application of competencies to ER Tohoku 2015
Systems-thinking competence	"An ability to collectively analyze complex systems across different domains (society, environment, economy, etc.) and across different scales (local to global)…"	An ability to systematically analyze interconnected issues of the reconstruction process (e.g. coastal structures, fishery industry, aging society, housing reconstruction) and categorize them across different domains (social, environmental, economic) and scales (individual, community, town, prefecture)
Anticipatory competence	"An ability to collectively analyze, evaluate, and craft rich "pictures" of the future related to sustainability issues and sustainability problem-solving frameworks"	An ability to envision, analyze and evaluate possible outcomes of different approaches to reconstruction process
Normative competence	"An ability to collectively map, specify, apply, reconcile, and negotiate sustainability values, principles, goals, and targets"	An ability to recognize and analyze different values that guide the decision making process of individuals and groups of people regarding reconstruction process An ability to take a normative standpoint when participating in decision-making process within a group
Strategic competence	"Ability to collectively design and implement interventions, transitions, and transformative governance strategies toward sustainability"	An ability to design and implement strategies for reconstruction process
Interpersonal competence	"An ability to motivate, enable, and facilitate collaborative and participatory sustainability research and problem solving"	An ability to contact, approach and communicate with various stakeholders in the field involved in reconstruction process An ability to work before, in and after the field in a multinational group of students and faculty members

The proposed key competencies were discussed among students during the unit to achieve a common understanding of them in the context of the activities and outcomes of the field exercise. The competencies were chosen as a framework for assessing the educational contribution of the unit in the context of the graduate program in sustainability. GPSS-GLI, however, neither utilizes the framework for any assessment within the program nor adopts the above-mentioned key competencies as a core for structuring the curriculum.

2.1 Description of Study Area

In September 2015, members of GPSS-GLI participated in the Tohoku Exercise on Resilience, which was centred on Otsuchi Town. The main outcome expected from the students was for them to understand the damage caused by the disaster and identify the issues that the affected area continues to face in its recovery process. In order to do this, participants attended lectures given by different town members, conducted interviews, toured local facilities, and took part in cultural exchanges with the local community, among other activities.

Nestled between the mountains and the sea, Otsuchi Town is located in Iwate Prefecture and faces two bays: Funakoshi Bay and Otsuchi Bay (see Figs. 2 and 3).

Fig. 2 Map of Otsuchi town, showing the location of Funakoshi Bay and Otsuchi Bay

Fig. 3 Otsuchi Bay from Machikata

Prior to the 2011 Tohoku Earthquake and Tsunami, the economy of Otsuchi was focused around the service sector, but with a significant contribution from salmon fishing, aquaculture of scallops and seaweed, and the fish processing industry (Esteban et al. 2015).

Otsuchi was home to around 16,000 people prior to March 11th, 2011. However, after the 9.0 magnitude earthquake and the subsequent tsunami the landscape was forever changed. Otsuchi was one of the hardest hit towns, due to its positioning along the rias coastline, the signature coastline topography of Iwate prefecture (Esteban et al. 2015). As a result of the tsunami 803 people died in Otsuchi, and 431 people are still missing, with an additional 50 related deaths recorded due to other indirect consequences, meaning that in total the disaster caused 1,284 victims (according to official data from April 1st, 2014, Otsuchi Town Government). As of August 2015, the town has a population of 12,452, representing a 22.1 % population decrease compared to the pre March 11th, 2011 baseline (Otsuchi Town Government 2015). In terms of physical damage, 451 ha of land were inundated, with 52 % of residential and 98 % of commercial areas flooded (Otsuchi Town Government 2015). Local industries were also significantly affected, with factories and service buildings washed away, fishing boats destroyed, and transportation networks cut (Esteban et al. 2015; Otsuchi Town Government 2015).

Currently, the town of Otsuchi is in the middle of the reconstruction process (see Figs. 4 and 5), following a three-stage reconstruction plan laid out by the local government. The first stage, referred to as "Foundational Reconstruction",

Fig. 4 Full view of the reconstruction process in Otsuchi town

Fig. 5 Ground elevation in Machikata

took place from 2011 to 2013, followed by the second stage, "Full-fledge Reconstruction", which was ongoing at the time of writing and is expected to be completed by 2016. The final stage of reconstruction is planned to take part between 2017 and 2018 and is termed as "Linking to Further Development" (Iwate Prefectural Government 2015). The reconstruction plan revolves around three pillars: ensuring safety, rebuilding lives, and rebooting industries. In the process of incorporating these pillars, Otsuchi Town is balancing several reconstruction issues, providing students of the field exercise an environment to learn the challenges of building resilience.

2.2 Level 1: Application of Educational and Research Field Methodologies

2.2.1 Educational and Research Field Methodologies in Exercise on Resilience, Tohoku Unit 2015

The group participating in Tohoku Unit was composed of three PhD and five Master's students, accompanied and guided by three faculty members from GPSS-GLI. The students represented a diverse group, with one Japanese student and seven international students. During the duration of the field exercise the faculty and the four Japanese-speaking students within the group served as translators in the communication with local residents and other key informants and town officials.

Prior to the field exercise, three lectures were provided by faculty in the period between mid-June and mid-July, which gave an overview of the 2011 Tohoku Earthquake and Tsunami and its consequences to the students. Past publications on natural disasters, management strategies, and the Tohoku region were distributed as supplementary reading material.

Various field educational and research methodologies were utilized to achieve a holistic understanding of the issues that the Tohoku region in Japan, and more specifically Otsuchi Town, is facing. Part of these methodologies were designed into the unit in order to give students both basic information as well as expert opinions. During the course of the unit, the students themselves implemented additional methodologies in order to reach a deeper understanding of the issues involved. This section lists such methodologies under the groupings, (1) Lectures, (2) Guided Tour, (3) Interviews, and (4) Other experiences.

(1) *Lectures*

Five lecture-style information sessions were scheduled over the course of the unit. Lectures with stakeholders that have direct contact with the study site provided first-hand information, data about the damage, and their own perspective on the ongoing issues that Otsuchi Town is facing. After each lecture students could ask questions in order to further clarify the issues.

The sessions included lectures from Officials from Otsuchi Town and from the Prefectural Government of Iwate that outlined their engagement and the progress in reconstruction. Members of a government-funded non-profit organization [Otsuchi Reconstruction Promotion Group (Otsuchi Fukko Suishin Tai)] working on the revival of the town shared their involvement and perspectives on the community and reconstruction efforts. One lecture with a professor from the Atmosphere and Ocean Research Institute (AORI), The University of Tokyo, introduced students to some of the issues and challenges facing the region. Finally, the head of the Sanriku Fisheries Research Center of Iwate University introduced the institution and gave his perspective on marine resources management and the outlook of fisheries in the area.

(2) *Guided Tours*

Guided tours of the disaster struck area and its surroundings allowed the students to better understand the context and scale of the physical damage that resulted from the earthquake and tsunami. The Field Exercise began in Otsuchi with a tour by a volunteer town guide, a key contact for the unit who had served as the district community head for twenty years and assisted the evacuation of community members at the time of the tsunami. His initial explanation of the town layout and evacuation process was followed by a tour of the evacuation route itself.

Another day was reserved to visit other communities along the Sanriku coastline, allowing students to see the different coastal defense strategies that were attempted by various towns. Also, it allowed students to see future infrastructural and zoning approaches that are being developed to defend against future calamities in the town of Taro, which could be contrasted with that of Otsuchi. Rikuzentakata showcased the failure of a forest as a protection countermeasure against the tsunami, as well as the sheer scale of land elevation projects (see also Esteban et al. 2015).

(3) *Interviews*

Interviews provided personal perspectives by Otsuchi residents and their own life stories, allowing students a more intimate connection with the residents of the town and the reconstruction process.

The first interview consisted of a visit to a seventy-year old business owner in the fishing industry who had lost his processing factory to the tsunami. He emotionally explained pictures of the disaster and discussed the difficulties of his business, as is expecting a season of unprecedentedly low catch, at the conclusion of a three-year federal subsidy, facing uncertainty regarding the building ground for a permanent factory.

Lastly, students were given the discretion to design and conduct a field survey for one afternoon. The group organized into smaller groups and toured the town by bicycle while conducted impromptu interviews on the streets and in shops, coming back together to exchange their findings in the evening. One of the groups spent time at the temporary housing unit they had visited previously, where they had a more in-depth conversation covering topics such as the living situation, community dynamics, and restoration efforts. Other groups engaged in conversations with shop owners, a priest, a security officer for construction, and a mother of a child with disabilities.

(4) *Other experiences*

Other experiences included a social gathering with residents of a temporary housing unit organized by the volunteer guide from the first day. A socially and culturally contextualized view of afflicted community members was gained from observing and interacting with the 14 temporary housing residents. The participants from the temporary housing unit ranged in age from around 68 to 94 years old.

In addition to the planned activities, group discussions and work in the evenings, shared meals and rooms, walking to activities and lecture venues instead of using public transportation offered students many opportunities to familiarize themselves with the town and reach a deeper understanding of the context and physical damage of the earthquake and disaster.

2.2.2 Identification of Sustainability Issues in Reconstruction Process

A total of 81 issues were identified by the participants of the Tohoku 2015 unit, which are listed in chronological order in Table 2 within the corresponding activity that aided their identification. From each activity carried out during the field exercise, students identified a number of issues related to the reconstruction process.

2.3 Level 2: Development of Competencies for Sustainability Research

During the exercise students developed a number of competencies that are essential for sustainability science (see Table 1). For example, during the interview session at the Fishery Processing Plant (see Table 2) it was mentioned that fishermen had experienced a drastic decrease in the salmon catch and they feared it could become a regular problem in the future. After identifying this issue, students incorporated this subject in their following activities, trying to gain a holistic understanding of the problems the fishery industry of the region is facing. For instance, other issues related to this problem were identified, such as the lack of labour force and successors in the fishing industry due to shrinking and aging population, damage to storage and processing facilities due to the tsunami, ecological reasons for salmon catch decrease and the problems of the Japan Fisheries Cooperatives (JF). The brainstorming and listing of the issues related to the fishery industry can be seen as an exercise encouraging system thinking. Anticipatory competence was practiced when discussing future scenarios for the fishing industry (with and

Table 2 Identified issues in reconstruction process, Tohoku Unit 2015

#	Activities	Identified issues
1	Tour with community head	Lack of water and toilet after evacuation/debate on land elevation plan/land ownership slows down the reconstruction process/community dynamics: Population drain
2	Government lectures: local (Otsuchi), prefectural (Iwate)	Depopulation/transportation: Temporary houses-downtown/ lacking participation of youth/dropping participation public forum/financial burden on people with pre-existing loans/ land with loan cannot be sold/local government carries maintenance cost of temporary housing/loss financial revenue from housing taxation/long negotiation between local and national government/lack of intergenerational dialogue
3	NPO (Otsuchi fukko suishin-tai)	Invisible wall between locals and newcomers/slow decision making: NPO—government/top-down approach
4	Dinner at temporary housing	Loss of cultural traditions (e.g. Matsuri)/emotional damage/ small temporary houses/no other future perspectives for living/lack of community gardens
5	Interview: Fishery Middle Man	Lower fish catches/too many activities proposals/temporary factory cannot meet safety regulations/depopulation is causing a lack of successors and human resources/reduced interest fishing industry/lack of building areas/tsunami tourism interfering with daily work/lack of diversification creates more risky situations
6	Visit to AORI, The University of Tokyo	Gap between research and helpfulness for residents/nature as an opponent/need for research apart from Tsunami in the ocean/only residences can rebuild on elevated land/need for a bigger port
7	Lecture: AORI Professor	Salmon juveniles not released on time. Result: Low fish catch/number of fishermen reducing while average age is going up/difficult to receive fishing permits as new fisherman/fish population reducing (e.g. illegal fishing, climate change, pollution)
8	Walk tour	Physical structures against the Tsunami failed/earthquake caused land subsidence/daily reminders of the Tsunami/car dependence in the community
9	Research satellite Iwate University	Climate change affecting water temperature: Salmon move North/unsustainable conventional fishing techniques/genetically homogeneous fish do not return to the river/education for fishermen/the initial investment cost for adapting aquaculture is high/lack of human resources in the fishery industry/difficulties to find the balance between pushing the university ideas and simply presenting them/mentality issues and indecisiveness within the town limiting the reconstruction speed/terminology barrier between the scientists, people, and fishermen/the architecture and building techniques need to be rethought considering the potential natural disasters/catching salmon in international waters causing the reduction of the number of Salmons

(continued)

Table 2 (continued)

#	Activities	Identified issues
10	Interview with community head	Disconnection among community members and people moving out/lack of mental health in the community/no enough numbers of public houses for all the applicants/ the houses are too small for families/quality issues of the temporary houses/difficulties for the new people to fit in the temporary housing communities
11	Interviews to local residents 1	Financial difficulties making people's life difficult/incomplete public transport system/emotional baggage due the difficulties of the life after the Tsunami/concern about the disconnection among towns by the sea walls/a disconnection between human resources and jobs/school facilities are needed for renewal
12	Interviews to local residents v2	Unclear work durations for worker/lack of entertainment within Otsuchi/few shops are available/difficulties to reclaim the same land as before/government has no plan for people to move out from the temporary housing unit
13	Bus tour	No standard for the coastal structures/local people's fear of the ocean/debate on the height of the seawall/no enough attraction for construction workers to stay in a long run/ ecological cost of cutting down the mountain and effects on the micro-climate/seawater inundation on land/a high cost to build the Tsunami wall with no co-benefit/the physical constraints making the life of local people difficult/potential lack of water infiltration/uncertainty for people moving back to Rikuzentakata city

without intervention). When interacting with different stakeholders in the fishing industry, students recognized diverse normative standpoints. During interviews with academia some strategies for tackling the low salmon catch were discussed. Finally, by interacting directly with stakeholders and by discussing among students and faculty, the interpersonal competence could also be developed through the analysis of this particular issue.

2.3.1 Assessment Methodology of Field Exercise Contribution to Competencies

Although the conceptual discussion about competencies relevant to sustainability is active in literature, attempts to empirically assess their acquisition are limited. For instance, Barth et al. (2007) presented findings of their qualitative analysis of three focus group discussions conducted among those enrolled in the Sustainability study program at the University of Lüneburg in Germany, although this study is based on a different framework for competencies.

In their paper, Wiek et al. (2011a) makes it clear that the scope of their study does not include a discussion on pedagogy of the competencies, or how to carry out a program evaluation based on them. In this section of the chapter, authors make an effort to analyze using a methodology that is developed by themselves in this chapter, the effectiveness of field exercise activities for obtaining the five key competencies introduced above. Such an analysis was carried out from a both quantitatively (questionnaire surveys among participants) and qualitatively (group discussions) point of view, as explained below.

Quantitative Analysis—Design of the Survey

Through a series of student group discussions, the authors formulated a questionnaire that aimed to measure the overall effectiveness of the Tohoku Exercise on Resilience in developing the competencies of the students that participated in it. Firstly, the framework for the assessment and the definition of each competency were explained to the students. A discussion followed on the five competencies and how various activities may have fostered them, where each student was asked to assess the unit's contribution to their personal development regarding these competencies. As the discussion had led to the distinction of passive and active styles of learning, the questionnaire items followed this categorization, asking students to self-evaluate their active and passive learning for each competency. Additionally, an item was added for each competency on how the unit led to their "recogniz[ing]/agree[ing] about the importance of the competency for research on sustainability issues." Each category of passive learning, active learning, and recognition of the importance of the concept was evaluated with a five-point Likert scale (1: very ineffective (No influence); 2: ineffective; 3: satisfactory; 4: effective; 5: very effective). Finally, a free response section prompted respondents for explanations of their answers and for relevant comments.

Passive and Active Learning

Pre-existing literature on the learning process is consistent with the understanding of active and passive learning that emerged in the student group discussions. Bonwell and Eison (1991) are still considered leaders on the topic (Michel et al. 2009) and note that the usage of the terms has been based on intuitive understandings rather than on technical, universal definitions. Various studies define their own understanding of "active" learning, for example as one with "fuller engagement with the material" (Benware and Deci 1984) or that "hold learners responsible for their own learning" (Cui 2013; Michel et al. 2009). A review of works on sustainability literacy found active learning, "self-directed enquiry, self-reflection, learning by doing, engagement with real-life issues and learning within communities of practice" to be a commonly understood requirement (Bergmann 2012). Noting the absence of a clear definitions of the terms, Chi (2009) proposed a framework and definitions for active, constructive and interactive learning. Although active learning was the shallowest of the three forms discussed, her paper acknowledges passive learning as a yet shallower form of learning. As this chapter only distinguishes between active and passive learning, it should be noted that all forms of learning that involve active engagement (including constructive and interactive processes) fall under the authors' understanding of "active" learning. On the other hand, "passive" learning may occur when information or knowledge is shared through lectures or other methodologies that do not necessarily require direct participation or engagement from the students.

Qualitative Analysis—Students' Group Discussion

Before conducting the questionnaire survey regarding the effectiveness of the unit for acquisition of the five competencies, a number of group discussions were held amongst students to ensure a mutual understanding of the competencies existing in the context of the field exercise. Group discussions were also found useful for linking the development of competencies to the activities within the unit, as well as obtaining qualitative data to complement the results of the questionnaire survey. This summary of group discussions does not represent a consensus, but rather highlights the various opinions exchanged by students during the sessions. The learning process may vary greatly by individual, and the summary illustrates a range of interpretations of quantitative survey results.

2.3.2 Assessment Results

Quantitative Data—Survey Results

The average responses of the eight student members of the Exercise unit are shown in Fig. 6. With the exception of active learning on the strategic competency (m = 1.63; stdev = 0.52), students indicated that the Exercise overall had been "satisfactory" to "effective" in facilitating the development of the respective capacities. Average scores across all three response categories were highest for interpersonal competency (m = 3.75), followed by anticipatory (m = 3.63), systems thinking (m = 3.54), normative (m = 3.51), and then by strategic competency (m = 2.72). Further, while the interpersonal competency was perceived as having developed more through active (m = 4; stdev = 1.07) than passive learning (m = 3.5; stdev = 1.2), on the whole, the evaluations indicated that students perceived more activities had a passive (m = 3.58) rather than an active learning approach (m = 3.08). The students also indicated that the Exercise had been effective in increasing their recognition of the importance of the competencies (m = 3.73).

Qualitative Data—Summary of Students' Group Discussion

A summary of the main results is presented in Table 3, showcasing the relation of some of the field methodologies within the design of the field exercise and the identification of issues, which derived from group discussions amongst the students. Additionally, the same figure relates the field methodologies to the respective competencies that each methodology is seen to promote more effectively. For instance, activities included in the category of lectures contributed to the identification of 21 social issues, 7 environmental issues, and 6 economic issues. The table shows, for example, how within the lectures the activity with Otsuchi Town Government would have mostly contributed to the development of the anticipatory and normative competencies. More details on students' discussions and this table are summarised in the following subsections.

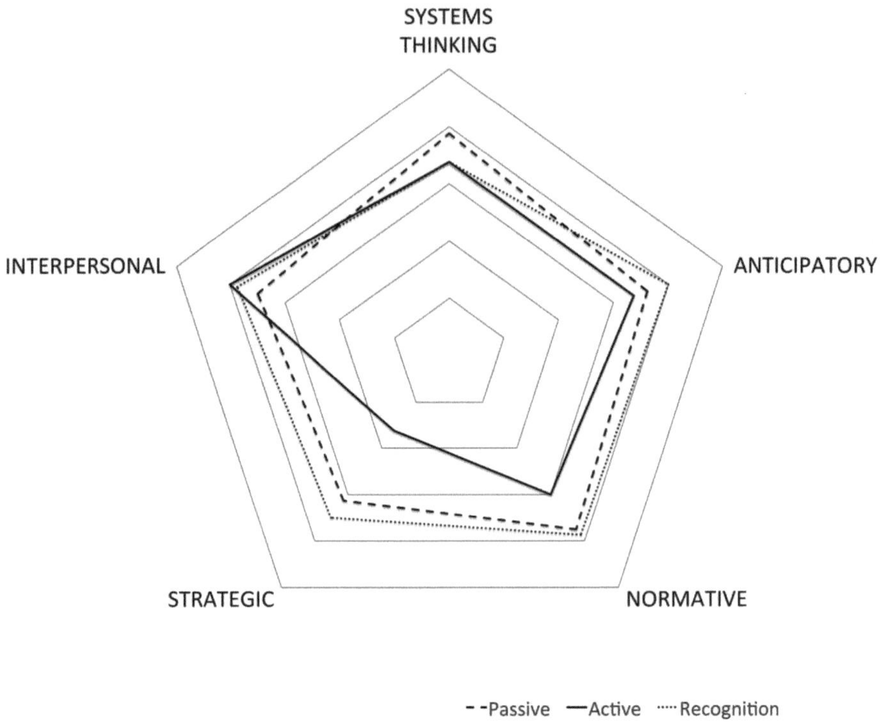

Fig. 6 Students' evaluation of competencies development during the ER Tohoku 2015

Table 3 ER Tohoku Unit 2015—field methodologies, results and competencies

Activities / Issues Identified	Lectures					Guided Tour			Interviews			Other experiences			
	Otsuchi Town Gov.	Iwate Pref. Gov.	NPO	AORI	Iwate Univ.	Tour with community head	Bus Tour	Walk Tour	Fishery middleman	Community head	Local residents	Dinner at temporary housing	Direct Observation	Group discussions	Living cond.
SOCIAL	Aging/Depopulation (2) Construction (2) Education (1) Decision Making (6) Research (4) Living Condition (2) Mental Condition (2) Fishing Industry (2)					Aging/Depopulation (1) Construction (3) Decision Making (2) Living Condition (3) Mental Condition (2) Land (1) Future Uncertainty (1)			Aging/Depopulation (1) Construction (2) Mental Condition (1) Fishing Industry (2) Land (1) Town Activities (3) Temporary Housing (6) Local Shops (1)			Education (1) Living Condition (5) Mental Condition (2) Sense of Disconnect (2)			
	TOTAL ISSUES: 21					TOTAL ISSUES: 13			TOTAL ISSUES: 17			TOTAL ISSUES: 10			
ENVIRONMENTAL	Salmon Fishing (6) Water Infiltration (1)					Change in Landscape (2) Water Infiltration (2)			Fish Populations (1)						
	TOTAL ISSUES: 7					TOTAL ISSUES: 4			TOTAL ISSUES: 1			TOTAL ISSUES: 0			
ECONOMIC	Loans (2) Housing (2) Port (1) [illegible]					Construction Costs (1)			Safety Regulations (1)						
	TOTAL ISSUES: 6					TOTAL ISSUES: 1			TOTAL ISSUES: 1			TOTAL ISSUES: 0			
Systems Thinking			□		□				□	□					
Anticipatory	□	□		□	□		□		□	□	□		□		
Normative	□			□	□	□	□			□				□	
Strategic					□				□	□					
Interpersonal										□	□	□		□	□

(i) System-thinking competence

The activities that took place during the unit were designed in a way that allowed students to learn about different stakeholders' opinions and views on the reconstruction process in Otsuchi Town and other locations of Iwate Prefecture coastal area. For example, the problems of the fishery industry in Otsuchi Town were introduced from different perspectives: ecological, social and business. For the case of the decline in salmon catch, scholars from Iwate University and The University of Tokyo AORI discussed the ecological reasons behind it during the lectures. The information session with the head of a temporary fishing processing facility fostered students' understanding of small and medium scale fishery businesses and the challenges they are facing. The lecture by the representative from the town government pointed out how the problems of a shrinking and aging population can affect the future of the fisheries by leaving fishermen in the region without successors. Moreover, the NPO (Otsuchi Fukko Suishin Tai) representatives gave an explanation of their collaboration projects with the fisheries for revitalizing the community. At the same time, the in-depth conversation with the head of the temporary processing facility mentioned above helped students realize the limitations of fishermen in being able to accept the offers from outsiders. Overall, communication with different stakeholders helped students to analyze the situation of fisheries from different angles and perspectives. Students had similar experience when exploring other issues such as house reconstruction, determining the height of seawalls and new land-use planning.

(ii) Anticipatory competence

The anticipatory competence is represented by the ability to foresee or identify future scenarios given a sustainability issue. During some of the activities in the field exercise the students were exposed to the objectives and planning of the reconstruction process, which represent a sustainability vision for the community. The students had a chance to recognize different perspectives for the future reconstruction process, and that the current reconstruction plan was structured for a selected future vision for the community. In this regard, the acquisition of an anticipatory competence through passive learning as well as the recognition of the importance of this competency was, on average, scored highly by students.

According to the group discussion, most of designed activities of the unit contributed to the development of anticipatory competence. For example, regarding the reconstruction issue, lectures by town and prefectural government representatives explained different approaches to the reconstruction process and shared future vision of the town and coastal area of Iwate Prefecture. The visit to the temporary houses, an in-depth interview with the community head (volunteer guide), as well as additional interviews conducted during field survey, gave a more personal perspective of the housing reconstruction issues. It also made students think about the future of each individual that makes the community. For example, the future of the fishing industry was discussed during the interview with the owner of temporary fishery processing facility as well as during Q&A sessions after the lectures with representatives of the academic community from Iwate University and AORI.

Although the average score for anticipatory competence was higher than satisfactory (m = 3.63), some students commented that they did not feel any increase in their anticipatory thinking compared to before the unit started. Moreover, as the design of the field exercise did not require students to propose a solution for the issues encountered, participants were not required to actively apply the anticipatory competence.

(iii) Normative competence

Through the activities carried out within the unit, different scenarios or sustainability visions for the community were presented by different stakeholders, and students were exposed to a range of value-oriented perspectives from different stakeholders involved in the reconstruction process. Choosing to rebuild the historical part of the town on elevated land over relocating, or choosing to move to higher inland areas over building a higher seawall represents decisions that are based on the values of the people making them, and not solely on technical aspects. The lecture by the Otsuchi Town government illustrated how these decisions varied even within one town, and the bus tour showcased different approaches for reconstruction in neighbouring cities. At the same time, the lectures by professors from Iwate University and The University of Tokyo AORI discussed different options for securing the future of the fisheries in Iwate prefecture, ranging from a reformation of the Fishermen Association to shifting almost entirely to aquaculture.

Some students recognized that having a chance to listen to various opinions on the same issue was effective for fostering their normative competence. Direct interaction with various stakeholders also contributed to students' understanding of their different circumstances and values. There were also comments from students stating that their normative competence did not increase much, as they were already used to actively utilizing it in their original field of study.

(iv) Strategic competence

Strategic competence was rated the lowest both during the group discussion and in the questionnaire survey. It was generally agreed among students that the development of strategic competency was one of the weakest parts of the unit, as there was no requirement to design possible solutions for the issues identified. However, some students mentioned how listening to the story of the owner of the fishery processing facility, as well as discussion with academy representatives about the future of the fishery industry in the region, provided a chance to exercise strategic thinking.

(v) Interpersonal competence

The group discussion indicates that there are different dimensions to the interpersonal competence that developed through the field exercise. The curriculum was built in a way that there was a collective participation in the core unit activities, such as lectures and interviews, as well as other experiences, such as transportation, room sharing and taking meals. In the evenings, group discussions and reflection sessions followed up on the day's activities. Students' group discussion results show that both the activities designed as well as living conditions and spending free time together during the unit were effective for developing communication and collaboration skills within the team. The structure of the unit required a considerable amount of group work and a number of collective decisions, which also contributed to fostering negotiation and consensus-building skills.

On the other hand, key-informant interview, lectures' Q&A sessions, as well as visiting the temporary housing units and having informal conversations with local residents, were also recognized as useful for the development of the interpersonal competency. As the activities designed were organized by faculty members, students were not required to contact the stakeholders directly prior to going to the field. However, as sustainability issues often include many stakeholders whose opinion should be taken into account, this competence could have been further improved by having more direct and frequent communication with potential interviewees, key-informants and experts.

3 Discussion of Assessment of Tohoku Unit 2015

The application of multiple educational and research methodologies in this field exercise allowed for the identification of the issues surrounding the reconstruction of Otsuchi Town. The combination of a variety of activities seems to have allowed students to identify a great variety of issues in the three dimensions of sustainability (General Assembly of the United Nations, 2012): social, environmental and economic (Table 3). From Table 3, it can be appreciated that the lectures seem to have contributed the most to identify a large number of issues (a total of 34 issues were identified through the lectures, while the interviews helped to identify 19 issues, the guided tours 18 issues and other activities the remaining 10, see Table 2). Additionally, although most of the issues identified were in the social dimension, it was through the lectures that students were able to identify the majority of issues belonging to the environmental and economic dimensions. However, discussions among students suggest that lectures provided a more passive learning experience, and it was through additional activities that some issues could be presented in a different light and a better understanding could be developed. For example, the issue of labour shortage was covered multiple times by the town government, the prefectural government, and the researchers at AORI. The town and prefectural government covered this issue from the angle of population recruitment and local economy, and it could be argued that this was repetitive. However, researchers from AORI looked at the same issue from the perspective of a research institute in need of new facilities and support. The semi-structured interview with the owner of a local fish processing company further aided the students in understanding this issue, by placing the broad concept of labour shortage into a real-world personal example.

The different environments in which the students interacted with stakeholders contributed to different styles of outcomes. The dinner party setting at the temporary housing unit allowed to contrast real life accounts with the lectures provided by the governments. While the lectures provided more on-target information, raw data, and a setting appropriate to raising specific academic questions, the dinner party brought to attention the social environment in which many of the reconstruction issues are grounded in and in which solutions must be feasible. The activity with the NPO, although also in a lecture style, again provided a different environment from the governmental lectures. The NPO lectures brought in the voices of outside people coming into Otsuchi after the disaster and starting projects to aid the recovery of the community. Their lectures were generally reflections on personal experiences, and this, along with the younger age of the informants, created a more relaxed environment.

Developing a closer relationship with stakeholders in the community was considered important in the students' discussion, and can be explained through the sequence of interactions that they had with the head of the temporary housing unit. The first meeting with the informant was upon arrival, before any of the lectures took place, with students not yet having had time to formulate specific questions regarding town reconstruction issues. The second meeting at the Temporary Housing Dinner allowed for casual interactions in which students gained a greater cultural understanding of the area, while the third meeting allowed for students to ask targeted questions that were developed throughout their stay in Otsuchi. Although the same informant took part in the guided tour, the temporary housing dinner, and in the semi-structured interview, the relationship between the informant and the students changed over time, allowing them to share different information and to reach deeper into the issues the community was facing.

The interview group sizes also made a difference on the outcome. Smaller group sizes were less intruding and allowed for a more comfortable setting in which students were able to observe and ask about more personal issues regarding daily lives and mental health issues. Building relationships is important not only for gaining information on current research, but are also crucial in continuing relationships for the sake of future research and Exercise on Resilience units. For instance, the head of the temporary housing unit enjoyed the multiple visits and is interested in setting up several small-group home visits, along with a group dinner, for students in future Exercise on Resilience units.

The different activities (see Table 3) also allowed for the issues faced by the communities that make up Otsuchi to be framed in different scales, ranging from the local to the regional scale. Activities such as walking around and interviewing townspeople depict a picture very specific to Otsuchi. The interviews uncovered personal experiences that may differ from case to case within the same town, let alone between different towns. On the other hand, the bus tour to neighboring towns of Taro and Rikuzentakata showed the bigger regional picture. Both of these towns took on a different reconstruction process than that of Otsuchi, and both of these towns face different issues in their reconstruction.

However, it is important to keep in mind some of the limitations of the analysis carried out. The distribution of issues seen in Table 3 suggests that the design of the unit and the research methodologies applied during the field exercise emphasize the social aspects of the reconstruction process. A greater number of issues were identified related to the social dimension, and the progression of activities seemed to deepen understanding about them, but contributed little to expand the students' knowledge regarding the problems that might exist in the economic and environmental dimensions. This outcome could relate to the specific purpose of the Exercise on Resilience to promote understanding of social resilience, and also to the particular interest of the students attending the field exercise. However, the contribution of the unit to students' working understanding of resilience is not necessarily captured in the competency assessment framework. In the design of the unit, resilience was defined as "the capacity of a system, enterprise, or a person to maintain its core purpose and integrity in the face of dramatically changed circumstances" (GPSS-GLI 2015). Tohoku Unit aims to nurture the student's understanding of resilience by exposing them to the issues that challenge creating a resilient society, and although the competencies assessed in this chapter may contribute to the understanding of resilience, the degree of contribution is unknown.

In spite of the limitations of the assessment regarding the contribution of the unit to the understanding of the concept of resilience, a quick review of the past Tohoku unit final reports reveals that different student groups have approached the issue of resilience in different ways. Some groups adopted the ecological definition of resilience, as a measure of the amount of disturbance that a system can take before it switched over from one equilibrium regime to another (Gunderson 2002). On the other hand, other groups focused on the importance of soft measures for disaster risk mitigation, such as the role of community networks, to improve resilience. Again, there were those who focused on defining pre-existing and emerging problems that need to be addressed to build a "resilient" future. Still others relied on the Resilience Assessment Framework laid out by the Resilience Alliance (2010) (see Fig. 7), in order to analyse the ongoing revitalization efforts in the town. By looking at Fig. 7, it is clear that the unit does contribute to the understanding of at least part of the concept of resilience, by directly exposing students to the identification of sustainability issues during the reconstruction process. Nonetheless, as mentioned before, the extent of the contribution to the understanding of the other aspects of resilience was not captured in this assessment.

The framework of key competencies (see Table 1) was adopted in order to guide the analysis of the educational outcome of this Exercise within a Sustainability Science program, though, as noted above, this framework was retrospectively applied. The questionnaire conducted amongst the students was a means of quantifying the exploration that emerged out of the discussions (see Fig. 6). As the aim was to document individuals' perceived learning experience, students were instructed to consider only the Exercise's *additional* impact on their personal competence. Additionally, while examples of active and passive learning were provided for each competence and the competence types were discussed at depth, most of the students were unfamiliar with the concepts prior to

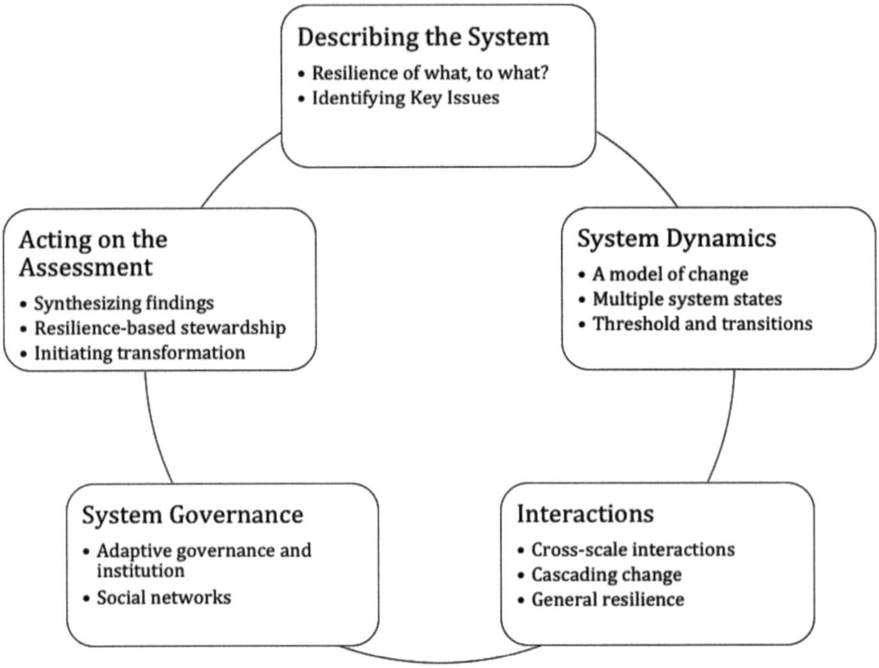

Fig. 7 Resilience Assessment Framework. Adapted from Resilience Assessment Framework (2010)

the Exercise, and the analysis relies on students' understandings of each concept. Results are thus subject to students' self-evaluative ability as well as to differences in pre-existing levels of competence and experience. Indeed, the discussions involved many instances of individual differences regarding whether or not certain activities were perceived to have contributed to certain competencies, and the group discussion results presented above do not necessarily represent a group consensus. However, ultimately the survey findings generally mirrored the results of the discussion.

As Wiek et al. (2011a, b) suggest themselves, their key competencies were not meant to be a finalized framework. Thus, the discussions and questionnaire have prompted the authors to think about the possibility of extending the typology and model described by Wiek et al. (2011a, b). The students agreed that the typology may be further developed, questioning, for example, whether the importance of the five competencies were really unique to sustainability science, or by suggesting "flexibility" as an additional key competency. Furthermore, students struggled to attribute activities to the development of specific competencies, and supported the notion that competencies were crosscutting.

4 Concluding Remarks

The Tohoku Unit focuses on the reconstruction process in Otsuchi Town, the sustainability issues that arise during this process and the characteristics of a resilient community in the face of natural disasters. In the light of the assessment conducted by the authors, some concluding remarks are summarized below:

- The proposed assessment framework is useful for the future design of field units, by identifying which activities contribute to specific dimensions of sustainability and by linking them with the competencies that they foster in students. Depending on the focus that faculty would like to set for future installments of a given unit, the design of the activities could be based on some of the findings of this chapter.
- The authors believe that the present chapter shows the importance of incorporating in the design of the unit a clear indication of the dimensions of sustainability and competencies that each field exercise aims to promote. The assessment made of the 2015 Tohoku unit highlights the imbalance among the dimensions of sustainability present in the unit. This could be explained by a conscious effort from the organizers of the unit to stress the importance of the social dimensions during the reconstruction process. However, it could be beneficial to restructure part of the unit in order to give more room to further highlight the environmental and economic dimensions in the reconstruction process.
- It is important for field exercises to break free from traditional modes of learning and be innovative about finding what is necessary for sustainability development to take place. As suggested by Benware and Deci (1984), students need to experience a more active paradigm of learning in order to recognize the passivity of traditional learning methods.
- Activities such as the tour and walking around the area did not lead to the discovery of a large number of issues or the development of competencies (see Table 3). However, they served the important purpose of experiencing places first-hand. It seems that it is indispensable to consider other factors not included in the framework, which could lead to identification of additional key competencies for sustainability. For instance, the role that people and their personal stories and experiences may play in sustainability education and research.
- The framework for key competencies seems to have been largely accepted by the students, who for the most part recognized their relevance to sustainability research (see Fig. 6). However, there are indications that these competencies might not account for the whole set of skills that students identified as necessary. For example, in the students' group discussion it was mentioned that having seen and experienced the field site (in its topography, spatial distribution, atmosphere, among other factors) allowed them to gain a deeper understanding of some of the previous knowledge they had received from the literature review and lectures. In the framework utilized in the present research it is not clear if this would belong to an existing competency or if it could be described through a new concept.

- A key limitation of the assessment framework utilized is the lack of depth in the analysis of the contribution of the unit to students' understanding of the concept of resilience. This is a limitation that should be address in future evaluations of this unit.

This reflective study offers insights into the capacity of Field Exercises such as the 2015 Tohoku unit to contribute to the personal development of students of sustainability science. Programs such as GPSS-GLI may consider the type of activities and unit structure that can be considered more conducive to the development of the competencies in future program designs. For example, much of the perceived interpersonal development was attributed to aspects of the Exercise other than the scheduled activities, and that may be reproducible in other settings. In contrast, the findings depicted the difficulties of fostering strategic competence under the current design. Organizers of the unit may consider the active development of strategic competences to be beyond the expected learning outcomes of an explorative field exercise and find other venues in the program for students' development in this regard, or to consciously target the current weakness in future designs of the unit.

Acknowledgments The authors would like to thank the Graduate Program in Sustainability Science—Global Leadership Initiative for providing the opportunity to participate in this field exercise and in the present book. In particular, the authors thank the faculty involved in the organization of the Tohoku Unit, Dr. Motoharu Onuki, Dr. Miguel Esteban and Ms. Izumi Ikeda. Acknowledgements go to the people of Otsuchi Town, Otsuchi Town Government and Iwate Prefectural Government, as well as to AORI Research Institute and Iwate University. Finally, the authors acknowledge the great contribution of Mr. Orlando Vargas and Mr. Yangqing Chen to the discussion during the assessment of this unit.

References

Ansari, W. E., & Stibbe, A. (2009). Public health and the environment: What skills for sustainability literacy—and why? *Sustainability, 1*, 425–440.

Baartman, L. K., Bastiaens, T. J., Kirschner, P. A., & van der Vleuten, C. P. (2007). Evaluating assessment quality in competence-based education: A qualitative comparison of two frameworks. *Educational Research Review, 2*(2), 114–129.

Barth, M., Godemann, J., Rieckmann, M., & Stoltenberg, U. (2007). Developing key competencies for sustainable development in higher education. *International Journal of Sustainability in Higher Education, 8*(4), 416–430.

Benware, C. A., & Deci, E. L. (1984). Quality of learning with an active versus passive motivational set. *American Educational Research Journal, 21*(4), 755–765.

Bergmann, I. (2012). The handbook of sustainability literacy: Skills for a changing world. *Environmental Education Research, 18*(5), 719–722.

Bonwell, C. C., & Eison, J. A. (1991). Active learning: Creating excitement in the classroom. 1991 ASHE-ERIC Higher Education Reports. ERIC Clearinghouse on Higher Education, The George Washington University, One Dupont Circle, Suite 630, Washington, DC 20036-1183.

Chi, M. T. (2009). Active-constructive-interactive: A conceptual framework for differentiating learning activities. *Topics in Cognitive Science, 1*(1), 73–105.

Cui, Y. (2013). An empirical study of learning outcomes based on active versus passive teaching styles. *International Journal of Education and Management Engineering, 1*, 39–43.
De Haan, G. (2006). The BLK '21'programme in Germany: A 'Gestaltungskompetenz'-based model for education for sustainable development. *Environmental Education Research, 12*(1), 19–32.
Esteban, M., Onuki, M., Ikeda, I., & Akiyama, T. (2015). Reconstruction following the 2011 Tohoku earthquake tsunami: Case study of Otsuchi Town in Iwate prefecture, Japan. Handbook of Coastal Disaster Mitigation for Engineers and Planners.
Estrada, E. (2014, July 28). Engineering Education: The U.S. Department of Education Releases Innovative Initiatives. Retrieved from Homeroom: The official blog of the US Department of Education. http://blog.ed.gov/2014/07/engineering-education-the-u-s-department-of-education-releases-innovative-initiatives/.
Frisk, E., & Larson, K. L. (2011). Educating for sustainability: Competencies & practices for transformative action. *Journal of Sustainability Education, 2*, 1–20.
General Assembly of United Nations (2012). The future we want. A/RES/66/288.
GPSS-GLI. (2015). Retrieved January 11, 2016 from http://www.sustainability.k.u-tokyo.ac.jp/.
Gunderson, L. H. (2002). Resilience and the behavior of large-scale systems (Vol. 60). Island Press.
International Commission on Education for Sustainable Development Practice. (2008). Report from the International Commission on Education for Sustainable Development Practice, 2008.
Iwate Prefectural Government, Coastal Region Development Bureau (2015, September) On the state of reconstruction from the Great East Japan Earthquake and Tsunami [PDF document].
Michel, N., Cater, J. J., & Varela, O. (2009). Active versus passive teaching styles: An empirical study of student learning outcomes. *Human Resource Development Quarterly, 20*(4), 397–418.
Mochizuki, Y., & Fadeeva, Z. (2010). Competences for sustainable development and sustainability: Significance and challenges for ESD. *International Journal of Sustainability in Higher Education, 11*(4), 391–403.
Otsuchi Town Government, Disaster Recovery Bureau, Recovery Measures Promotion Office. (2015). Status of Rebuilding Efforts in Otsuchi-cho, Iwate Prefecture [PDF document].
Resilience Alliance. (2010). Assessing resilience in social-ecological systems: Workbook for practitioners.
Rowe, D. (2007). Education for a sustainable future. SCIENCE-NEW YORK THEN WASHINGTON-, 317(5836), 323.
Segalas, J., Ferrer-Balas, D., Svanström, M., Lundqvist, U., & Mulder, K. F. (2009). What has to be learnt for sustainability? A comparison of bachelor engineering education competences at three European universities. *Sustainability Science, 4*(1), 17–27.
Shephard, K. (2008) Higher education for sustainability: Seeking affective learning outcomes. *International Journal of Sustainability in Higher Education, 9*(1), 87–98.
Voorhees, R. A. (2001). Competency-Based learning models: A necessary future. *New directions for institutional research, 2001*(110), 5–13.
Wiek, A., Withycombe, L., & Redman, C. L. (2011a). Key competencies in sustainability: A reference framework for academic program development. *Sustainability Science, 6*(2), 203–218.
Wiek, A., Withycombe, L., Redman, C., & Mills, S. B. (2011b). Moving forward on competence in sustainability research and problem solving. *Environment, 53*(2), 3–13.
Willard, M., Wiedmeyer, C., Warren Flint, R., Weedon, J. S., Woodward, R., & Feldman, I., et al. (2010). The sustainability professional: 2010 competency survey report. *Environmental Quality Management, 20*(1), 49–83.

Drawing Lessons from the Minamata Incident for the General Public: Exercise on Resilience, Minamata Unit AY2014

Eri Amasawa, Heng Yi Teah, Joanne Yu Ting Khew, Izumi Ikeda and Motoharu Onuki

Abstract This chapter introduces the Minamata Unit, a GPSS-GLI Exercise on Resilience conducted around Minamata City, Kumamoto Prefecture, in Japan. The disruption caused by Minamata mercury pollution persists even 60 years after the incidents. Such a long history of development can provide invaluable insights into the study of the impact of industrial pollution to a society, which includes the significance of identifying new pollutants, the intricate social tension between victims and local community, and the urge to heal the strained bonds. Drawing lessons from the Minamata incident can prevent and mitigate similar incidents in developing countries, where environmental regulation is typically less stringent than in developed countries. In the present chapter, the authors describe the outcomes of a week-long field exercise in Minamata, designed for students of sustainability science. The aims of the unit were for participants to understand the complexities of the Minamata incident, not only the causal relationship between the pollutant and its impact on human health, but also the impacts the disease had on the society. Also, it attempted to examine the responsibility of scientists and the government, especially when certain issues regarding the case remain scientifically uncertain, by redefining the issue in the contemporary context. In order to facilitate learning by the participants of the unit presented in this chapter and students of future units, the exercise organisers requested the production of educational materials as the final output. Students were divided into three working groups, each tasked with the development of a different type of media material, namely blog posting, video production, and game development. The prearranged field activities included stakeholder interviews, site visits, and an intensive group work. Overall, the student groups completed the production of the tangible outputs, though their effectiveness in reaching the target audience and helping future sustainability science students require further analysis.

E. Amasawa (✉) · H.Y. Teah · J.Y.T. Khew · I. Ikeda · M. Onuki
Graduate Program in Sustainability Science-Global Leadership Initiative,
Graduate School of Frontier Sciences, The University of Tokyo,
5-1-5 Kashiwanoha, Kashiwa, Chiba 277-8563, Japan
e-mail: aamasawa@s.k.u-tokyo.ac.jp

M. Esteban et al. (eds.), *Sustainability Science: Field Methods and Exercises*,
DOI 10.1007/978-3-319-32930-7_5

Keywords Pollution disease · Dissemination of information · Communication skills

1 Introduction

The present chapter introduces the Minamata Unit, a GPSS-GLI Exercise on Resilience (ER) conducted around Minamata City, Kumamoto Prefecture, in Japan. As summarised in Chap. 1 of the present book, ER have been designed as a specific part of the GPSS-GLI curriculum, focusing on giving students a fieldwork experience that would contribute to their understanding of the concept applied to real issues in sustainability. The ER is predominantly focused on education, rather than research. In the one-week period, the Exercise attempts to concentrate the attention of participants on understanding the complexity of sustainability issues in the field, and thus their core purpose is for students to be exposed to the issues of concern.

GPSS-GLI first organised the ER Minamata Unit in the Academic Year (AY) 2011. The ER Minamata Unit has always aimed at providing students with the opportunities to gain a comprehensive idea about the Minamata incident with an emphasis on the social aspects of the issue, and to examine the responsibilities of each stakeholder, namely those of scientists and government. It is important to note that many issues remain scientifically uncertain, and the unit attempts to redefine the issue in a contemporary context.

The present chapter summarises the approach of the Minamata Unit in AY 2014, highlighting the methodology used to facilitate students' understanding, as well as to develop skills to communicate the knowledge gained by students from the field exercise to other sustainability science students and practitioners. Many approaches can be taken to achieve the goals and produce an output from exercise courses. The 2014 exercise adopted a collection of creative group projects as its core internal dynamic, with a focus on education for sustainability. The objective of the group projects was for participants to understand the various perspectives of the Minamata disease sufficiently to produce a deliverable output that could communicate the ideas and experiences gained to the wider public, sustainability science students, and practitioners. Many of the topics and problems associated with the disease continue to influence Japanese society at present, but people's understanding of the risks and scientific knowledge remain limited.

The approach brought about another advantage that is as equally significant as the learning experience itself: the improvement of the communication skills of the participants. Encouraging participants to contemplate and ultimately to summarize their experience in a tangible form does not only further their understanding, but also improves the skills required to become a sustainability scientist. Especially with the issue of the Minamata disease, participants are likely to go through a rather painful emotional turmoil of disturbance, agitation, anger, and sorrow.

Being able to organize their thoughts under such circumstances, and to be able to achieve a holistic understanding of the issues involved in a scientific yet simple manner, is the most critical vision of the Minamata exercise.

Given the complexity of the issues involved, the authors will first provide a brief account of the history and development of the various issues regarding the Minamata disease, which served as a background for much of the dissemination material that was produced as a result of the field exercise.

1.1 Minamata Disease and Mercury Pollution

The disruption caused to the human settlements around Minamata city due to mercury pollution persist even 60 years after the incident took place. The Minamata disease refers to the symptoms exhibited as a result of mercury poisoning, which was caused by the release of water-borne methyl mercury into the environment (Harada 1995). As the disease mostly affected residents in the vicinity of Minamata City in Kumamoto Prefecture (see Fig. 1), it has become known as the "Minamata disease". Sufferers of the Minamata disease experience symptoms similar to those of the Hunter-Russel syndrome, which were originally identified following research on mercury poisoning that resulted from the consumption of seed preservatives in England during the 1930s. Hunter-Russel symptoms include: sensory disorder, ataxia, visual field constriction, impaired hearing, and speech impairment. However, thousands of sufferers have experienced a range of milder symptoms which are still being exhibited today. The accurate diagnosis of sufferers has proven difficult, resulting in a multitude of legal, social, political, and medical disputes over the last 60 years.

The culprit for the mercury pollution was the Chisso Cooperation, a major fertilizer manufacturer in Japan (Minamata City 2007a, b). In its early years the Nippon Nitrogen Fertilizer Corporation, the predecessor of Chisso Corporation, established in 1908, utilized the mountains close to Minamata City to generate hydroelectric power to for their energy-intensive production process. Mercury was only used at Chisso from 1932 when mercury oxide was introduced as a catalyst in the synthesis of acetaldehyde. However, the mercury oxide catalyst is reduced in the process of acetaldehyde production. For continued acetaldehyde production, oxidation of the reduced catalyst was carried out using permanganate (MnO_4^-), in a method specially devised by Chisso. This method was later modified in 1951 to use Ferric Iron (Fe^{3+}) and nitric acid as a replacement for permanganate. Although this change was suspected to be the cause of the onset of the Minamata disease in 1953, scholars were unable to accurately identify the source of mercury pollution, which is now understood to have originated from the wastewater released into Minamata Bay from Chisso's acetaldehyde production process. Chisso also did not assist in investigations and attempted to deflect their responsibility (see Table 1 for a timeline of the most important events in this period). In their defense, Chisso raised two main objections: If they were

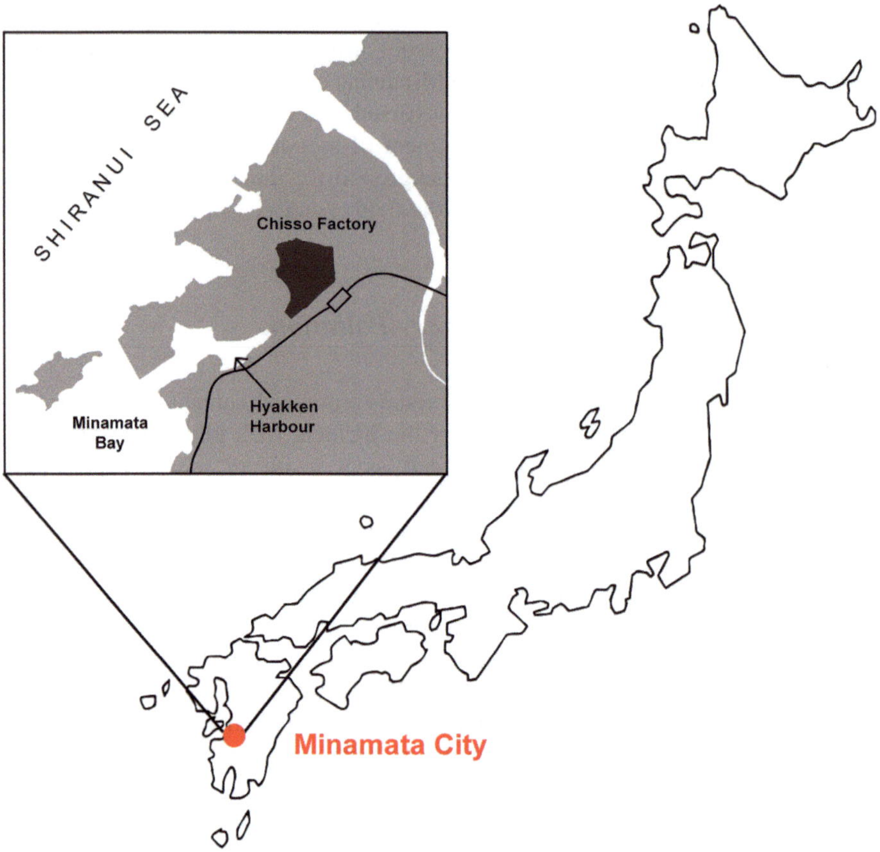

Fig. 1 Map showing the location of Minamata city, Kumamoto Prefecture, Japan. *Inset map* shows the location of the Chisso Cooperation's factory, and its wastewater discharge point, Hyakken Harbour (*Image* created by the authors using information from the "Map of Minamata Bay and the Chisso Factory", Wikipedia, CC BY-SA 3.0)

truly the culprits, (1) why did victims of the Minamata disease only appear in the 1950s, although they started the acetaldehyde synthesis process in the 1930s; and (2) why were victims only found in the vicinity of Minamata city, although acetaldehyde synthesis was carried out in factories around the world? (Nishimura 2006)

1.2 Examining the Cause of the Minamata Disease

Around the 1950s/1960s, during the time when it was scientifically proven that the Minamata disease was caused by water-borne mercury, Chisso was aware that

Table 1 Timeline of important events concerning Chisso factory and mercury pollution (Minamata City 2007a, b; Nishimura 2006)

Year	Event
1908	Chisso inaugurated in Minamata Inorganic fixation of nitrogen gas (N_2) was proceeded for use in fertilizer production
1930	Inorganic synthesis of ammonia (NH_3) from nitrogen gas (N_2) and hydrogen (H_2) gas was proceeded for use in fertilizer production
1932	**Organic synthesis of Acetic acid** Acetylene ⟶ Acetaldehyde ⟶ Acetic acid *Catalyst:* Mercury sulphate / chloride Permanganate (VII) ion (MnO_4^-) was used as a catalyst oxidant in Chisso's unique method Wastewater from factory processes was discharged into the Minamata Bay
1951	Chisso changed its original process for the organic synthesis of acetic acid by replacing the permanganate (VII) ion (MnO_4^-) with the more common method of utilizing nitric acid (HNO_3) as a catalyst oxidant This method resulted in the production of methyl mercury from the mercury sulphate/chloride catalyst Wastewater from the factory process was still discharged into Minamata Bay
1953	Start of the Minamata Disease outbreak
1956	Official report and confirmation of Minamata disease
1958	Chisso changed the route of the acetaldehyde effluent channel from Hyakken port to Hachiman pool; effluents were then discharged into the mouth of Minamata river
1959	Under government orders, Chisso stopped discharging wastewater to Minamata river Instead, they claimed to have redirected it back to Hyakken port through an improved wastewater treatment facility
1962	First official identification of congenital cases of the Minamata disease
1966	Chisso changed its effluent process system with the introduction of a complete acetaldehyde effluent processing circulation system Discharge of effluent that contains methyl mercury was halted
1968	May: Chisso halted production of acetaldehyde September: The national government officially pronounced Chisso factory as the responsible party for the outbreak of Minamata disease
1969	Legal decree set to prevent the industrial sector from discharging effluent that contains methyl mercury (under the Water Quality Control Act)

their wastewater contained traces of mercury. Indeed, Chisso authorized a medical doctor of its corporate hospital to feed food that was contaminated using wastewater to cats in secret, resulting in cat #400 developing symptoms typical of the Minamata disease (dis-coordination).[1] Despite such knowledge, Chisso refused to acknowledge evidences of Chisso's responsibility in causing the Minamata disease outbreak in 1953. According to research by Nishimura (2006), in the early 1930s

[1]Today, a tombstone commemorating the death of the cats used in Chisso's wastewater toxicity experiment stands outside Soshisha's Minamata Disease Museum as a permanent reminder of Chisso's responsibility in the Minamata disease incident.

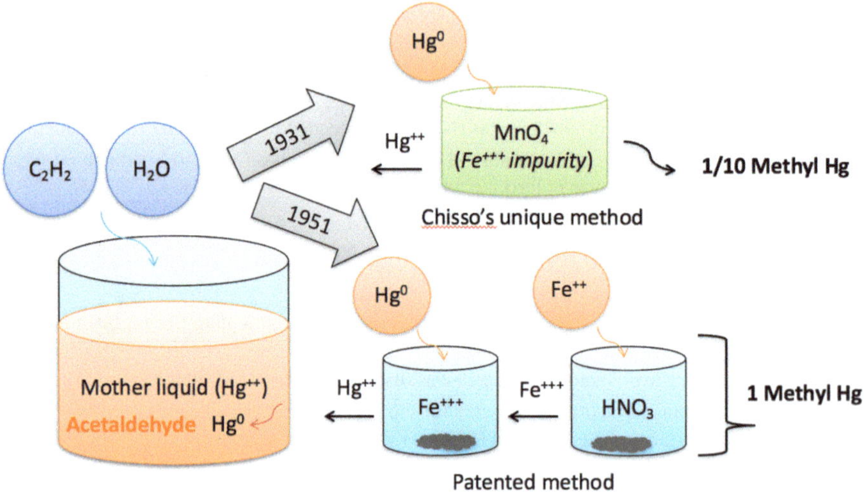

Fig. 2 The two different acetaldehyde production processes used by Chisso in 1931 and 1951, respectively. The change from Chisso's unique method to the patented method increased the levels of methyl mercury discharge, explaining the onset of Minamata disease

Chisso initially utilized their own process for oxidizing the mercury oxide catalyst (reduced during the process of acetaldehyde synthesis) in order to avoid paying the high patent cost attached to the prevalent commercial method. Chisso's process involved permanganate as an oxidizing agent, while the prevalent commercial method uses ferric iron. However, the reason behind the effectiveness of Chisso's method was ironically due to the presence of ferric iron impurities in the permanganate oxidizing agent. Chisso's method also coincidentally lowered the production of methyl mercury by about one tenth of what would have been produced if the commercial method had been used. During the post-war rebuilding years, the production of materials such as polyvinyl chloride was important, and Chisso switched from their method of acetaldehyde synthesis to the established commercial method, which utilized ferric iron, in order to increase production output. This method required the use of another oxidizing agent (nitric acid), to oxidize the reduced ferric iron (Fe^{2+}) back to Fe^{3+}, producing even more methyl mercury in the process (see Fig. 2).

In retrospect, it is important to note that the Minamata disease did not only occur in Minamata City. In 1965, methyl mercury was also released by a factory into Agano River in Niigata prefecture, resulting in people around its vicinity exhibiting a range of mercury intoxication symptoms (Funabashi 2006). However, what was different in the case of the methyl mercury discharge of Showa Denko (Niigata Prefecture) and that of Chisso (Minamata City) was that the latter was situated nearer to the sea and hence wastewater was discharged directly into the seawater. According to Nishimura (2006), the concentration of chlorine

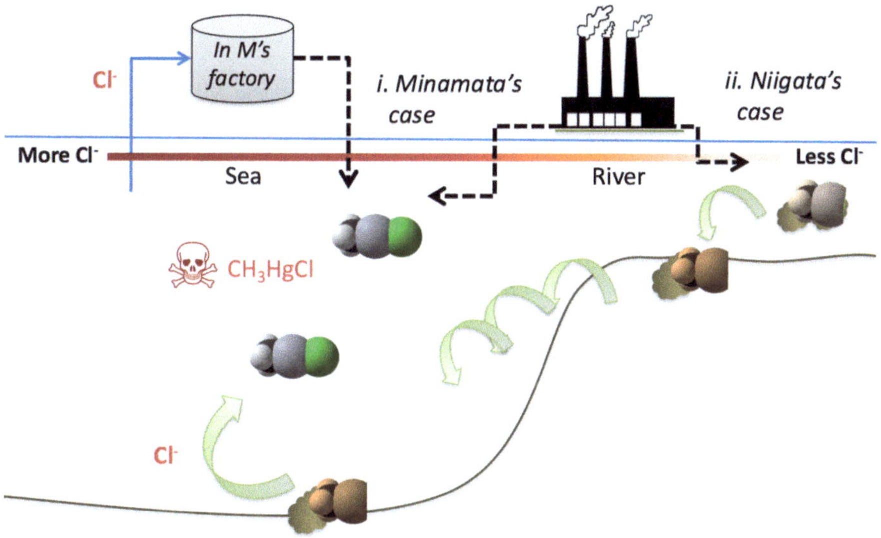

Fig. 3 Difference in the release of methyl mercury iron into the sea or into a river

ion present in seawater would affect the solubility of the methyl mercury compound. When wastewater containing methyl mercury is discharged directly into the seawater (as in Chisso's case), the high concentration of chlorine ions attached to mercury form soluble methyl mercury chloride compound. This compound is highly mobile in water and can be easily taken into the body tissue of sea-creatures and humans that directly or indirectly ingest it. In the case of mercury pollution in Niigata, mercury-containing wastewater was discharged into a river where the concentration of chlorine ion is much lower than those in seawater. The methyl mercury ion then conjugates with abundant organic debris present in the river water (e.g. protein colloids, algae) and sinks into the riverbed. As methyl mercury of this form is insoluble and therefore not mobile, it is not as readily transferred to the human body. However, mobility of this form of methyl mercury can occur when it is consumed by benthic organisms that feed or live on sediment; and biologically magnified up the food chain; or when it is transported to the sea along with riverbed sediment through erosion (Fig. 3). Once at the sea, the methyl mercury ion trapped in the insoluble compound recombined with the abundant supply of chlorine ions in the seawater, forming soluble methyl mercury Chloride. In Chisso's case, seawater was also used for the cooling of factory reactors, resulting in the formation of more methyl mercury chloride. As such, the two arguments posed by Chisso in its defence were eventually rejected. However, it should be noted that these explanations were reached only after decades of research.

1.3 The Spread of Minamata Disease

Once methyl mercury from Chisso's wastewater entered the Minamata bay in soluble form, pollution spread throughout the bay. As industrial waste was released through freshwater, the direction and magnitude of spread was determined by the salinity of the seawater that it was discharged into. In addition, the methyl mercury was taken up through the gills of oceanic organisms or through the ingestion of contaminated prey. It is important to note that methyl mercury persists in the tissue of marine organisms for about 72 days, and is biologically magnified up the food-chain. Hence, the movement of fish within Minamata bay further widened the range of area that was polluted. Minamata bay is situated within the Shiranui sea, which is bordered by many islands and landforms, creating a rather closed system with a prominently north to south current. Fish tend to gather and traverse the Shiranui sea using the main current and thus form the bulk of the catch of the fishermen residing in the surrounding islands and lands. Consumption of mercury-contaminated fish thus contributed to the spread of the pollution to areas bordering the Shiranui sea.

This geographical expansion of the area affected complicated the issues, especially regarding its influence on the local community. Delay in the establishment of access to the relief system in neighbouring municipalities, a regional gap in the level of medical care and compensation, and other issues further compounded the problem. Even nowadays, only a fraction of the victims on the islands bordering the Shiranui sea are officially certified as Minamata disease patients, leaving the support for "forgotten victims" largely to several non-governmental organizations active in the area.

1.4 The Status of Compensation, Legal Aid, and Political Settlements

Since the outset of the Minamata disease lawsuits have been filed for over 50 years, which have generated a variety of governmental compensation and relief schemes (Minamata City 2007a, b). In 1959, researchers from the Kumamoto university first published a proof that wastewater from Chisso's factory was the main agent behind the Minamata disease, communicating their findings with those of the experiments performed on cats (Nishimura 2006). However, Chisso refused to acknowledge their fault and insisted on making payments as "Mimaikin (sympathy money)" instead of a full compensation to victims. The criteria for victims to receive "sympathy money" from Chisso under the "Mimaikin Solatium Agreements" were unfair (weighted in Chisso's favour), and could be paid only to those approved by Chisso. However, official lawsuits started in 1969, one year after Chisso was found to be responsible for pollution by the national government (in 1968). The Kumamoto district court eventually ruled in favour of the plaintiff in 1973.

Meanwhile, the governmental assistance scheme, under the Law Concerning Special Measures for the Relief of Pollution-Related Health Damage, was promulgated in 1969. This scheme, in combination with the 1973 verdict, resulted in an increased number of applications filed for certification as patients. However, this change caused a severe delay in the reviewing process for certification. Accelerating the processing of applications proved a serious problem for the prefectural government (Minamata City 2007a, b). In response, the national government announced new certification criteria with stricter conditions for certification in 1977.

However, there was a string of parallel lawsuits filed by victims who were rejected by the certification board and their supporters. In 1995, in order to ease the mounting tensions, the national government made a political settlement that aimed for the national and prefectural government, as well as Chisso, to provide lifetime medical compensation. This compensation is offered based on the severity of the sufferer's symptoms, regardless of their certification status. Furthermore, in 2004 the Supreme Court, for the first time in the history of Minamata disease lawsuits, recognized the responsibility of the national and prefectural government (the Kansai Lawsuit). This verdict led to new legislation in 2009 to further relax the criteria for compensation and medical assistance. Spanning several decades, these cumulative legal fights (which are still continuing today), have prevented many victims from receiving compensation. Despite the first and second Minamata disease lawsuits and their corresponding compensation and relief schemes, there are still a large number of patients who remain unaccounted for. This is due to the difficulty in ascertaining their symptoms, as some of them are similar to those of old age and thus generate problems with the categorization of their compensation level.

1.5 Industrial Pollution Impact on the Environment and Society

Societal conflicts between various citizen groups have emerged as a result of the Minamata disease incident. Even as early as 1908, before Chisso's predecessor company established their factory in Minamata city, societal rifts were present between the fishermen in the south and the other citizens who reside closer to the city centre. Due to their status as migrants (from other parts of the country), and their low income, people from the fishing communities were seen as lower-class citizens. Chisso's activation of factory production of acetaldehyde in Minamata City in 1930 brought economic growth to Minamata city, which provided resources for the city to expand, supplied jobs, and increased living standards. When the Minamata disease started to spread, the people who were most affected by mercury poisoning were the fishermen—the group of people who benefited least from Chisso's economic contributions to Minamata city. In the early stage of the incident, while the cause was still unknown, the disease was called the “strange disease”. Patients of the Minamata disease were ostracized by other citizens due to the misconception that the disease was contagious. When Chisso was found guilty of causing the Minamata disease in 1968, many Minamata City citizens were employees of the factory. In such situation, Minamata City citizens had little incentive to be sympathetic towards the victims of the Minamata disease, partially in fear that they may lose their job for being regarded as a hostile to the employer, or that the company would suffer economically if they were ordered to grant high compensation. Also, sufferers of the Minamata disease became hesitant to apply for certification in fear of being discriminated against, and fishermen were scared of being banned from the market and thus lose their income. There were also conflicts among the patients who received compensation and medical aid, because of the differences in perceived fairness of the demands being made, due to the varying symptoms and reflected amount of compensation. Perhaps the most ironic of all developments is that Chisso still remains the major economic provider of Minamata city today (Minamata City 2009). Despite causing the Minamata disease, participants of the Minamata Unit found through interviews with local residents that Chisso is still regarded as the employer-of-choice.

The social problems that plagued the islands surrounding the Shiranui Sea still remain 60 years after the onset of the Minamata disease. One of the lessons from the Minamata disease saga is that pollution is not merely just an environmental issue. The discharge of pollutants into the Minamata Bay could have been swiftly halted by the industry. However, the choice to favour economic benefit over human health resulted not only in huge environmental problems but also in social conflicts and discrimination. Ideally, in order to promote sustainable development, industrial pollution should be entirely avoided, though this intent appears difficult for the case of developing countries (Li et al. 2009).

The Minamata city government marks every 1st of May as the commemoration day of the Minamata disease, and in recent years, the city has actively rewarded environmental activists locally and globally, and promoted environmental clean-up programs (Minamata City 2007a, b). Minamata city now has Japan’s most comprehensive recycling program, with 22 different ways of classifying garbage. The city has also been selected in recent years as Japan’s greenest city (Minamata City, 2015). These steps have helped to improve the image of the city, though it still suffers from the past shadow of the Minamata disease saga.

2 Methodology

In order to maximize understanding of the Minamata disease, participating students were divided into groups, where each group was assigned to produce a creative output that could transfer their learning to the public and students in future Minamata Unit. Unlike typical academic dissemination methods (i.e.

conferences, papers, and seminars) that target professional or technical crowd, the general public was chosen as an audience to challenge students to express Minamata issues using rather simple language. Consequently, dissemination media were chosen based on resource accessibility and facility of production in a short period of time, namely blog posts, a video clip, and an educational game. Additionally, the term ‘public’ here was not limited to Japan but encompasses audiences in other countries, and thus outputs were produced mainly in English.

2.1 Exercise Structure

The RE Minamata was structured into three phases that included preparation, field visits, and post-field visit activities, with students having to complete assignments in each of the phases. The role of each phase is explained in the following subsections and summarized in Table 2.

2.1.1 Preparation

After the formation of groups, the exercise theme, group work objectives, and schedules of field visit sites were communicated to students during an introductory orientation session. Students were divided into three groups of four to five members, where each group was assigned to produce one unique output from the three dissemination methods: blog posts, a video clip, and an educational game. Ideas and resources for the creation of each type of media were introduced by the academic staff responsible for the exercise, but each group ultimately established a detailed structure and contents of their project by themselves.

In order to help students organize their projects, each group was asked to prepare and submit a 3-page proposal. The proposal template was adapted from the United States National Science Foundation fellowship application. Faculties and staff responsible for the exercise unit reviewed the proposals submitted, and a feedback session was held a few days before the departure to the field.

In terms of academic lectures, students were required to take a prerequisite course that provided an overview of Minamata disease prior to participating in the exercise. Additionally, a professor in Public Health presented scientific background on mercury poisoning.

2.1.2 Field Visits

The field visits for the AY 2014-15 unit were conducted from February 27th to March 4th, 2015. Students visited museums and historic sites relevant to

Table 2 Role and details of ER Minamata Unit structure

Phase	Timeline	Agenda and purpose	Time spent during official activities[a]	Assignments due
Preparation	20 days before departure	Group formation Lecture on Mercury and Health	1.5 h	N/A
	2 days before departure	Feedback on proposals	2 h	Group work proposal
Field visit	Day 1	Kumamoto Prefectural Government Office Minamata Disease Municipal Museum	Day-long field visits and 2 ~ 3 h of group work time	N/A
	Day 2	Soshisha • Story telling by patients • Tour of Minamata disease museum • Guided tour of Minamata city (former Chisso factory's wastewater discharging point, nearby fishing villages, etc.)		
	Day 3	Minamata Disease Shiranui Patients' Association Minamata Disease Patients Alliance (Interviews conducted in Gosho-no-ura Island)		
	Day 4	Minamata Disease Victims Mutual Aid Association and the Collaboration Centre for Minamata Disease Victims in Minamata Minamata High School		
	Day 5	Lecture on history and lessons learned from Minamata disease by the director of division of Minamata Disease, Kumamoto Prefectural Government Exchange of ideas with young professionals of Minamata and its nearby area		
	Day 6	Summary of field visit		15 min oral presentation per group
Post field visit activities	1 month after field visit	Completion of output for publishing on the internet	5 h	Presentation on output and submission of deliverables

[a]This time does not include the amount of time that students spent on private study and group discussions before and after the field visit

Fig. 4 Visit and interview with one patient support group

Minamata disease, and rest of the time was spent on learning the ideas of different groups, organizations, and individuals through listening to storytelling, talk series, and panel discussions (see Fig. 4). The unit organizers selected these interviewees in advance to reflect the diverse perspectives and complex relations among local stakeholders. As a result, the field program included interactions with governmental officials of Kumamoto prefectural government, representatives of four different patient groups/patients supporting groups, academia (in the field of process engineering), high school students, and local young professionals.

After dinner each day, there were several hours that were designated for group work. During such periods, students organized the information they obtained through the interviews and other activities on the field, and discussed with faculty and other resource persons to synthesize the knowledge gained into their final output. The last day of the field visit period was designated for presentation of the output.[2]

[2]At this stage students were not required to show the final output, but to explain the structure it would take and the progress that had been made so far.

2.1.3 Post Field Visit Activities

Students continued working in groups for approximately one month after the field exercise to finalize their output, which was required to be of a quality high enough that it could be published on the Internet. The official last day of the exercise was April 10, 2015, when all members gathered and each group presented their output.

3 Results: Process and Outcome from Group Projects

3.1 Blog Posts

3.1.1 Goal and Scope

The blog group aimed to holistically disseminate the lessons learned from the Minamata incident. The central idea was that industrial pollution was not a mere environmental problem, but its causes and impacts were associated with the effectiveness of governance, scientific knowledge, local economy, and the people. To create public awareness, the objective was to publish a concise blog that covered all those aspects. In order to do so, a cartoon sketches (or figures) were produced to improve the reader friendliness of each post. The blog was written in English and Chinese, in order to reach out to wider audiences (see Fig. 5)

3.1.2 Materials and Methods

The materials for writing the blog were primarily based on the prearranged lectures (Table 3) and interviews during the field visits. Additional materials included The Science of Minamata Disease, a book written by Professor Nishimura that described scientific challenges in identifying the pollution source, and Minamata Disease—Its History and Lessons, an official document published by the Government of Minamata City (Minamata City 2007a, b) to depict the outlook of the city's future.

The blog was prepared using two different sets of processes. First, a blog post was written the evening of each day for the entire duration of the field visits, archiving the group's activities and the impressions gained from each visit in a chronological manner. After the field visits, a thematic blog was written using materials from the chronological blog, synthesizing findings that covered different aspects of industrial pollution. It began with an introduction to Minamata disease, a timeline of incidents, the scientific challenges in identifying the pollutants, the spread of disease, the compensations and settlements, and the societal conflict. Both of the blogs were published on the GPSS-GLI student website.

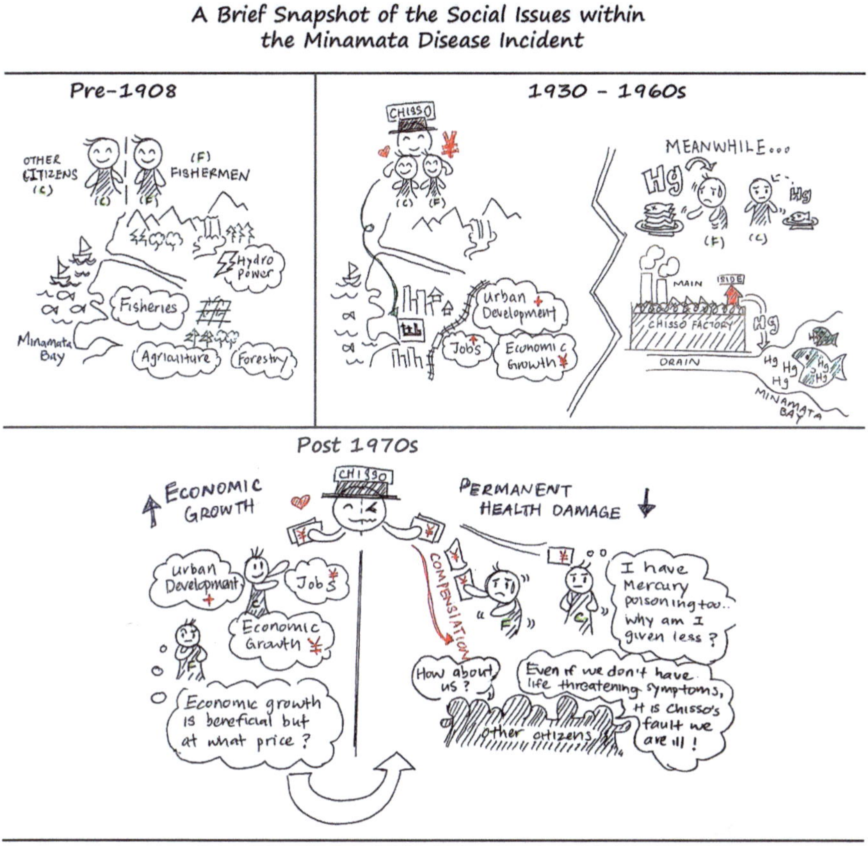

Fig. 5 Chisso's involvement in altering the social conflict scene in Minamata City. http://st.sustainability.k.u-tokyo.ac.jp/2015/04/22/an-introduction-to-the-minamata-disease-7 l societal-conflicts/

3.1.3 Results

The group published a complete blog entitled "Minamata Story in 8 Posts" on May 1st, 2015 (which is the official memorial day of the Minamata incidents).[3] Each post is about 300 words, and contained one original figure/sketch produced by students. The titles and main messages of each of the posts are summarized in Table 3.

The cartoons or figures presented in the blog post were considered to be helpful to communicate the outcomes and messages from the exercise. For example, Fig. 2 shows the two different acetaldehyde production processes which were

[3]The blog is accessible at the URL: http://st.sustainability.k.u-tokyo.ac.jp/category/minamata-unit-2015/.

Table 3 The summary of the titles and main messages of the "Minamata Story in 8 Posts" blog

	Title	Main messages
1.	What is the Minamata disease?	Minamata disease is a sickness caused by methyl mercury poisoning. The symptoms include sensory disorder. Less acute symptoms are hard to distinguish from other general aging afflictions
2.	Chisso Factory and Mercury Pollution	Chisso Factory is the culprit of the mercury pollution. A change of production method in 1951 triggered the disease
3.	Examining the causes of the Minamata disease (1)	The volume of released methyl mercury increased markedly after 1951
4.	Examining the causes of the Minamata disease (2)	A similar production was applied elsewhere but the impact was much smaller (as the chloride concentration in the water body into which the pollutants were released was different)
5.	The spread of wastewater, biomagnification and fishermen's catch	The disease spread to a wide area despite the fact that pollutants were discharged into Minamata bay only, because it accumulated in fish and other sea animals
6.	The status of compensation and political settlements	Lawsuits requesting relief measures or compensation persisted for more than 50 years. Different schemes for compensation and political settlements were provided to the victims
7.	Societal conflicts	The community was divided into victims and defenders of Chisso (because they were dependent on the company for jobs). Also, Minamata residents were stigmatized as disease bearers/carriers
8.	Environmental pollution or societal problems	Industrial pollution is not a mere environmental problem. The impact on society is tremendous and costly

implemented by Chisso Factory in 1931 and 1951 (presented in the third blog post, "Examining the Cause of Minamata disease"). The latter process produced ten times more methyl mercury, which helps to explain why the disease was diagnosed only in the 1950s, as explained earlier. Figure 3 shows the societal conflict scene involving Chisso in the period between the 1930s and 1960s and after the 1970s (presented in the seventh blog post, "Societal Conflicts"). Chisso promoted urban development and job opportunities in the city, though this largely ignored the fishermen, despite the fact that they were the first to be impacted by the pollutants accumulated in sea products. The lengthy compensation lawsuits further divided the community.

In short, the eight blog posts concisely described the Minamata incident in a holistic manner, which is a key feature of the analysis methods of sustainability scientists. Readers can grasp the importance of governance while scientific evidence lags behind facts on the ground, and the significance of the societal impacts that may result from industrial pollution.

3.2 Video Production

3.2.1 Goal and Scope

The goal of the video production team was to create a short documentary film that explained the current state of the Minamata disease to the world. The team decided to focus on three prevailing issues of the Minamata disease and a variety of stakeholders' perspective on how to resolve them. The three issues were the complexity of patient and/or sufferer certification process, clinical problems with the disease, and social discrimination within the community. The target audience of the film was university students with a basic understanding of the concept of sustainability, and members of the general public with an interest in Minamata disease. Prior knowledge of issues regarding Minamata was not necessary, as the film includes an introduction to the Minamata disease.

3.2.2 Materials and Methods

Basic equipment for filming was supplied by The University of Tokyo, which included a voice recorder, a hand-held video recorder, and a tripod (see Fig. 6). Film editing

Fig. 6 Recording of information and discussion session with one patient support group

was completed using a laptop, including software such as Adobe Premier, Windows Movie Maker, and GIMP. During the field visit the video team filmed all activities where permission was given to do so, and conducted few interviews with relevant stakeholders. Faculty and staff members negotiated with the representatives of all locations filmed and people interviewed in order to obtain filming permission. The contents as well as other materials presented in the film were carefully reviewed to avoid any potential violation of copyright. Since the spoken language is Japanese in Minamata, English subtitles were added to all Japanese conversations.

3.2.3 Results

The final output of the project was a 20-min film. The film begins with a brief introduction on the historical background of Minamata disease, and highlights current projects and issues raised by the various stakeholders, as well as their vision for the future of Minamata. The team explored and analysed the reasons behind the delayed settlement of the three issues through their interactions with the stakeholders, and presented recommendations on how to improve the current situation.

3.3 Game Development: Finding a Solution for the Issue of Compensation Regarding Minamata Disease

3.3.1 Goal and Scope

The objective of the game development group was to create an interactive game for graduate students to play in a classroom setting, which could be tested and implemented during a Minamata lecture in a Sustainability Science course. The game was assumed to take place after one introductory lecture on issues related to the Minamata disease, and thus players were assumed to have a basic understanding of the disease and its problems.

3.3.2 Materials and Methods

The game development team went through a great deal of exploration of educational games and how to integrate concepts used in such games into issues related to the Minamata disease, as they were given the freedom to develop any type of educational game. The team began their efforts by studying the Minamata disease through a literature survey. An early idea was to develop a role playing game in the hope of being able to convey the array of different stakeholder's thoughts and opinions. Although role playing can be an effective method to grasp the past and perceive the emotions of stakeholders through acting, it was argued that it could be inappropriate due to the sensitivity of the issues involved and ongoing lawsuits.

After travelling Minamata and interacting with local residents, the game group decided to shift their focus to the issues surrounding compensation and political settlements. Compensation and settlements for industrial pollution can result in a protracted conflict between a corporation and the affected population, but the case of Minamata is further complicated due to the extensive scale of the affected area and delayed response by national, prefectural, and municipal governments. Such a situation makes the settlements of Minamata disease a compelling topic for students of sustainability science to discuss.

3.3.3 Results

The developed game was named as "Let's find a solution for the compensation issue of Minamata disease victims." The overarching objective of the game is to experience the challenge of building a consensus regarding the formulation of Minamata disease compensation schemes.

To play the game, five players represent different stakeholder groups: the national government, patient groups, Chisso Corporation, social welfare groups, and citizens of Japan. Players are provided with a handout that details the background, situation and condition of each stakeholder. Then, all stakeholders are asked to focus on one issue of the Minamata disease relief scheme, namely the lack of a compensation program for the sufferers (i.e. uncertified patients) who are still experiencing physiological disorders. Example types of solution could include the provision of a medical notebook for unsatisfied patients, with benefits covering hospital expenses and commuting allowance, where Chisso Corporation and the national government agree to cover the expense for the compensation. Each player is given a number of opportunities to state their opinions and negotiate. At the end of the game, each player ranks the solutions presented by each stakeholder, and the stakeholder with a solution receiving the highest rank becomes the winner.

As a result of playing the game, participants are expected to gain an understanding of the current status of Minamata compensation and relief scheme, practice negotiation skills, and eventually apply their learning to other similar issues in the world. As of the time of writing this chapter, there has yet to be a chance to play this game, but all the instructions has been uploaded onto the same page as the blog website and it is expected to be be played by students of subsequent Minamata Unit.

4 Discussion and Conclusions

The ER Minamata Unit AY 2014 was the first attempt at GPSS-GLI to assign creative projects as outputs that participants had to deliver. The reason why creative projects were adopted was that although Minamata disease has a history of 60 years, with many issues still affecting society today, the understanding of such scientific issues by the larger society remains insufficient. Through this unit, students were expected to contribute to the society by disseminating valuable lessons learned. Since much research on the Minamata disease has already been carried out from a number of different research angles, studying them is much more important than attempting to develop new research questions. This is in direct contrast to many other field exercise units in GPSS-GLI, were students are encouraged to identify their own research interests and conduct surveys. Also, much active research is still being carried out on Minamata, and there are still politically delicate concerns and disputes. The participants of the Minamata Unit have to be conscious of the sensibilities of any stakeholders with a poor understanding of the background of the situation, and the organizers had to carefully prepare reading materials and arrange resource persons. Care was taken to make sure that, in situations where there were conflicting points of view, the resource persons and literature provided represented all existing points of view. As a result of such exposure, students were able to create original educational materials and experienced an effective learning environment.

Through the expansion of the scope of the exercise to dissemination of information to those outside of the academic community, the creative projects provided students with an opportunity to depict the incident in a comprehensive manner that reflected a variety of points of view. This allowed participants to deepen their understanding over the complex nature of the issue, and produce three different types of information materials for dissemination: blog posts, a video clip, and an interactive game for education. The true originality of the unit is thus not to develop a new research perspective, but to integrate existing perspectives into a coherent explanation of the many viewpoints and issues regarding this disease.

As the creative projects aimed to provide an opportunity for students to disseminate what they had learned to a wider audience, participants were challenged to produce outputs that were expressed in a simple and concise language for the general public. The blog team did this by using simple writing and by producing cartoon-like images that could be easily understood. The video production group identified three key issues that still persist in Minamata and visually portrayed what they learned. Finally, the game group composed an educational game that tackled issues related to compensation and relief schemes, which do require some prior knowledge of the issue. All three projects can help a member of the public or students of sustainability science to understand the current status of Minamata disease at the local scale, and simultaneously grasping the issue's universal importance. None of these projects could have been completed without a field visit to the actual site, and requesting such type of outputs from students was highly original from the point of view of GPSS-GLI.

Following the exercise unit, a short survey was conducted amongst the participating students (9 responses out of 11 participants were received). Overall, students perceived creative projects as an intellectually stimulating yet challenging activity that they would recommend other exercise courses to adopt. About one third of the students wished that they had more time for group work during the exercise, and one commented about the need for longer preparation period prior to the visit. The limitations of the use of creative projects as an educational tool appear to be similar to other types of field exercise outputs. Extensive literature review is always recommended prior to the visit, but learning from a book can never replace the physical experience of visiting a site. Students should be aware of the importance of flexibility in altering project content, and how the reality in the field is often different from their first thoughts and impressions after reviewing literature.

Acknowledgements The authors would like to express gratitude to blog post team members, Mahdi Ikhlayel and Angeli Guadalupe, for allowing the blog contents to be used in this chapter. Also, the support of Professor Hiroyuki Katayama and Dr. Jatuwat Sangsanont in organizing this unit is kindly appreciated. Finally, Mr. Yoshito Tanaka and Ms. Michi Nagano provided invaluable information and a basis for discussion from local perspectives. The authors would also would like to express special gratitude to Professor Emeritus Hajime Nishimura.

References

Harada, M. (1995). Minamata disease: Methylmercury poisoning in Japan caused by environmental pollution. *Critical Reviews in Toxicology, 25*(1), 1–24.

Minamata City. (2007a). Minamata disease: Its history and lessons. http://www.minamata195651.jp/pdf/kyoukun_en/kyoukun_eng_all.pdf. Accessed on 25 December 2015.

Minamata City. (2007b). http://www.city.minamata.lg.jp/Material/4662.pdf. Accessed on 3 January 2016.

Minamata City. (2015). http://www.city.minamata.lg.jp/1000.html. Accessed on 3 January 2016.

Nishimura, H. (2006) The science of Minamata disease in Japanese 西村肇,水俣病の科学,日本評論社.

Funabashi, H. (2006). Minamata disease and environmental governance. *International Journal of Japanese Sociology, 15*(1), 7–25.

Li, P., et al. (2009). Mercury pollution in Asia: A review of the contaminated sites. *Journal of Hazardous Materials*, *168*(2), 591–601.

Part IV
Sustainability Science Field Research

Sustainability Science as the Next Step in Urban Planning and Design

Giles Bruno Sioen, Toru Terada and Makoto Yokohari

Abstract The urban planning and design disciplines have repeatedly failed to build sustainable communities that are economically, environmentally, and socially viable and resilient. Sustainability science has the potential to be combined with the fields of urban planning and design, which primarily focus on the physical shape of the city, to develop new methodologies for building sustainable communities. To verify this, the present chapter aims to explore potential overlaps by identifying the field methodologies and focus of urban planners and designers, and that of sustainability scientists, through a multifaceted literature review. The narrative review carried out identified that methodologies applied within contemporary urban planning and design are not suitable to incorporate and solve underlying urban issues such as inequality or gentrification. The causes for this are likely related to the fundamental limitations present in urban planning, which has evolved from architecture, design, and engineering backgrounds that tend to have a specific vision of development predominantly dealing with design aspects and a focus on hard infrastructure. To overcome this issue, the authors discuss the potential role that sustainability science could play in opening up the field of urban planning and making it deal with underlying issues through the implementation of mixed methodologies (such as spatial analysis techniques, participatory tools, and qualitative or quantitative surveys) that can capture both the scientific reality and the contextual situation. Such mixed methods can provide a field researcher with broad problem identification tools, rather than focusing on specific physical and mostly morphological elements. In addition, the application of sustainability science could provide evidence for urban planning and design juries, inhabitants, and

G.B. Sioen (✉)
Graduate Program in Sustainability Science—Global Leadership Initiative, The University of Tokyo, Tokyo, Japan
e-mail: gilessioen@s.k.u-tokyo.ac.jp

T. Terada · M. Yokohari
Department of Urban Engineering, The University of Tokyo, Tokyo, Japan

M. Esteban et al. (eds.), *Sustainability Science: Field Methods and Exercises*,
DOI 10.1007/978-3-319-32930-7_6

decision makers to make calculated long-term decisions. Essentially, the present chapter argues that sustainability science can shift the methodologies used within planning and design towards the use of scientifically-oriented methodologies that help decision-makers create sustainable communities.

Keywords Sustainability science • Urban planning • Morphology • Mixed methodology • Public participation

1 Introduction

Sustainable communities are communities that are socially, economically, and environmentally viable (UN-HABITAT 2009). While sustainable cities are promoted as an ideal goal within a variety of policy contexts, questions arise with regards to which cities can address the challenges of sustainability (Bulkeley and Betsill 2005).

The field of urban planning should be responsible for promoting the viability of cities from a sustainable point of view. Urban planning emerged during the last century in Europe and the USA, closely linked to urban modernist ideas (Watson 2009) that including outdated ideas and promoted formalist physical solutions (Fainstein 2000). An example of such formalist physical solutions includes separate zoning, which separates each function within a city into different zones. This results in the separation of residential, commercial, and industrial areas instead of promoting a mixed use of functions (Yokohari et al. 2008).

However, it has become increasingly clear that a number of different planning concepts such as compact city long advocated by planners cannot be achieved merely by physical hard planning (Albrechts 2004). Different ideas and concepts are leading to change, but there is still no clear understanding of the effects that these ideas about the physical space have on social outcomes. Hence, questions regarding the appropriateness of urban designs are continuously being brought into debate (Fainstein 2000).

At this point, it is also clear that there is a need for planners to redefine sustainability, as its current formulation romanticizes our past and is too vague (Campbell 1996). Essentially, building sustainable communities seems to be for wealthy, developed nations and the focus on the morphology of the city has led to mere city beautification and marketing (Al-Soliman 1988; Moulaert 2004) over establishing good city functions and ensuring social wellbeing.

In this sense, it has also been argued that the physical planning and design of cities is incapable of dealing with the major urban challenges of the 21st century: resource depletion, rapid urbanization, poverty, and informality (Tibaijuka 2009). All these challenges are likely to be exacerbated by climate change, which is the biggest long-term challenge humanity is currently facing (IPCC AR5 2014; Mahlstein et al. 2013; Giddens 2009). Future changes in rainfall patterns may create regional water scarcity, which can be exacerbated by demographic trends such as on-going urbanization and large-scale agriculture (Jiang 2009; IPCC AR5 2014). Climate change may also have an added impact on the increasing older demographic profiles of many developed countries, as elderly people are more vulnerable to heat waves (Changnon et al. 1996).

Apart from decisions on specific adaptation strategies,[1] urban planning and design does not usually tackle the root of problems causing climate change, nor does it factor in the increasing speed with which impacts are expected to take place. The real issues that climate change will create, often discussed using terms such as Sustainable Development Goals (SDG), reside in factors that are beyond the physical shape of the community and therefore require different solutions. However, it is difficult to tackle such problems within the current paradigms of urban planning, essentially based on modernistic planning theories, of which the fixed physical shape of cities created from such theories clearly lacks flexibility. Given the challenges posed by climate change, with regard to communities need to adapt, and this adaptation to change is often described in terms of the concept of resilience.

The purpose of this chapter is to discuss a new role for urban planning in addressing the challenges of adapting to climate change and improving the resilience of communities. To do so, and if it is to achieve positive change, urban planning must move beyond only looking at physical solutions and integrate ideas and methods from the problem-solving approach that characterizes sustainability science (Miller 2013). In fact, the fields of urban planning and design have already made a start at motivating and provoking scholars and practitioners from various other disciplines. They have started to seek urban forms that will meet the requirements of sustainability, while enabling existing environments to function in a more constructive way than at present (Jabareen 2006). Many of the methods which urban planners use to analyze and understand urban areas have been analyzed from human geographers' attempts to conceptualize, differentiate, define, and interpret urban social, economic, and environmental change in the city (Dodson and Gleeson 2009). Thus, in order to achieve truly sustainable communities, sustainability science can help urban planners integrate mixed types of field techniques and public participation methodologies that go beyond the morphological, typological, and other physical aspects of urban planning and design that are commonly applied.

2 Methodology

This study was conducted through an analysis of secondary data, firstly through a review (Green et al. 2006) of interviews with renowned architects, urban planners and designers, and publications from international organizations that include specific field methods; and secondly by reviewing sustainability science journal

[1]Such as for example by increasing the height of seawalls to protect against storm surges in coastal regions.

papers. Such a mixed use of sources makes the methodology of this study unique in sustainability science research. The overall aim was to link the academic field of sustainability science with practical urban design and planning concepts that grasp the actual situation in the field and focus on identifying gaps between academia, policy, and the private sector.

The first keyword search that was used to filter relevant literature from the *Science Direct* search engine included the terms "urban design", "urban planning", and "sustainability". The papers identified through this process were then selected based on whether they featured prominent people from the 20th and 21st century who are relevant to the topic of this chapter. Identification of concepts and the overall focus of discussions in the field of sustainability science was conducted using a selection of journal papers based on the keywords "Sustainability science". The study was thus carried out in three main steps:

1. A review of narrative literature to understand how and why urban planning and design is focused on morphology, based on books, literature on interviews, and non-peer-reviewed literature and commentaries on prominent projects by renowned professionals.
2. A categorization of the results of the initial phase of the review into two groups: (1) urban planning and design focused on morphology, and (2) alternative concepts that solve underlying issues.
3. The final step in the literature review was conducted on scientific papers containing field methodologies in sustainability science focusing on the results they achieved.

The results of the literature survey were then discussed from the point of view of climate change, the three dimensions of sustainability (Lozano 2008) and improving the understanding of planners and designers about new ways to organize urban areas. Although this methodology is limited to literature involving urban planning, design concepts, and sustainability science, the study can provide a conceptual idea of how these different disciplines develop and complement each other over time.

3 Results

3.1 Trends of Urban Planning and Design

Urban planning and design draws on several field methodologies. Morphological research is key to understanding the city. Ethnographer Lévi-Strauss described the city as "*the most complex of human interventions (…) at the confluence of nature and artifact's. The city is the accumulation and the integration of many individual and small group actions self-governed by cultural traditions and shaped by social and economic forces over time. Urban morphologists focus on the tangible results of these forces: they study the outcomes of ideas and intentions as they take shape on the ground and mold our cities. Buildings, gardens, streets, parks, and monuments, are among the main elements of morphological analysis*" (Moudon 1997). The main method described by Lévi-Strauss as morphological analysis shows a focus on physical aspects over the interrelationships in the city. It is thus clear that urban planning and design cannot consider all interlinked aspects necessary to create a sustainable community (Moudon 1997).

The modernist movement led by Le Corbusier and the organization Congrès Internationaux d'Architecture Moderne (CIAM), dominated the 20th century and has shaped many contemporary cities (Madanipour 1999). The modernist design that flourished from 1928 to 1960, receives much credit for its rational modifications to the structure of cities and laid down the foundation of contemporary urban planning and design, posing a tremendous influence on the shape of cities and towns all over the world (Mumford 2002). The reasons behind the changes that these architects made to urban landscapes were often rooted in a lack of good planning systems, overpopulation, and the spread of diseases. A radical approach to planning has been necessary to solve the problems of that time. However, criticism towards the modernist movement include how they not only failed in achieving their objectives, but actually reinforced the patterns that they wished to change (Holston 1989). Basically, cities based on the modernist ideas have been recognized as inappropriate urban forms that can create and/or promote urban poverty and exclusion. Therefore, urban planning must be revised to create better livelihoods and promote inclusiveness and better environments for the poor (UN-HABITAT 2009).

In order to do so, it is imperative that architects and urban planners work together, yet they tend to disagree over many issues. The need for disciplines to co-exist and to foster interdisciplinary learning would appear to be obvious (Hirt and Luescher 2007), yet individuals appear to be destroying rather than complementing one another. Christian De Portzamparc, a renowned architects/planner in France and winner of the Pritzker Prize in 1994 said: "*The break between urban planning and architecture must be rethought today. The urban vision is often too technocratic. It ignores the spatiality. On the other hand, we can make cities with directors who know nothing of the actors.*" He also acknowledges that there should be another discipline that brings them closer and collaborate: "*We need to invent a new profession that is at the interface of these two disciplines. Also it's a dangerous job and the theoretical basis for practice does not exist. In fact, there is a real cry of urban thinking. This characterizes our time. So we must first develop a permanent research that makes investigations to feed the fieldwork both in terms of urban form, the volumetric, as programs*" (de Portzamparc 1995). Within this statement, de Portzamparc is speaking out as an architect criticizing the contemporary urban paradigm and calling for a new profession to have them collaborate.

Manuel de Solà-Morales (1939–2012), recognized for his minimalistic answers to sometimes-complex design situations, came up with the idea of "urban acupuncture". Essentially, Solà-Morales, idea was to make specific adjustments to the morphological shape of the city (Mitchell 2009). He came to the conclusion that a more interdisciplinary understanding of architecture was necessary, and that methods from different disciplines should be used to improve the concepts and methodologies that are traditionally used within urban planning and design. As a result, he initiated several changes in urban planning and design philosophy, while almost everywhere in Europe the Anglo-Saxon model of planning acquisitions in relation to the methodology of social sciences was applied (Smets 1994). Slightly later, Proctor (1998) proposed additional concepts (such as *ethics* and *morality*) that should also be taken into account, in addition to the physical urban forms that geographers were already addressing. This shows that ethics were rarely thoroughly considered within the discipline (Proctor 1998). As Eisenman (Hendrix 2006) and Proctor (1998) stated, methods that look at informal settlements or social informal constructs have already been recognized. However their implementation within the urban planning and design discipline is limited.

Kevin Lynch, a prominent American urban planner, stated: "*It is clear that the form of a city or of a metropolis will not exhibit some gigantic, static, stratified order. It will be a complicated pattern, continuous and whole, yet intricate and mobile. It must be plastic to the perceptual habits of thousands of citizens, open-ended to change of functions and meaning, receptive to the formation of new imagery. It must invite its viewers to explore the world.*" The field methodology he developed helps analyse the city by using five elements: "paths, edges, districts, nodes, and landmarks." (Lynch 1960). Lynch's obsession with form continued over a decade, from the publication of his first book in 1960, *The Image of the City* to the publication of *Good City Form* in 1984 (Lynch 1984). There, he talks about the ideal form, created through various spatial characteristics which should be: "General, linked with culture and basic values, dimensions of performance, connectable, measurable, independent, transformable over time" (Lynch 1984). Lynch was certainly not the only planner concerned with the urban form. Cedric Price condensed urban evolution into a metaphor of three different types of egg cooking: (1) The ancient city is a boiled egg, (2) industry city is a poached egg, and (3) modern city is scrambled eggs (Carmona 2010). The indication and focus of these morphological changes can be considered as the starting point of city analysis.

In conclusion, it is clear that some positive changes in the methodological approaches employed by urban planners and architects have been made. However, research remains largely organized within a given context, often limited to a specific country (Freestone 2014), and it is clear that it is still necessary to seek an alternative that will allow urban planners and architects to truly move beyond traditional methodologies in order to design more sustainable urban spaces.

3.2 What Alternatives Are Out There?

In order to identify promising alternative concepts and paradigms within urban planning and design it is necessary to categorize what has been already proposed in literature. The authors carried out this categorization according to the emphasis and focus placed by each of the concepts in literature identified on a given

Table 1 Categorization of concepts in urban planning according to authors

Category	Concepts	Source
Environmental	Ecosystem-based adaptation	Maes and Jacobs (2015)
	Eco-city	Roseland (1997)
	Landscape urbanism	Waldheim (2006)
	Garden cities of to-morrow	Howard and Osborn (1965)
	Ecological urbanism	Mostafavi and Doherty (2010)
Social	Machizukuri	Sorensen (2002)
	Yonmenkaigi	Na et al. (2009)
	The death and life of great American cities	Jacobs (1961)
Economic	Levittown	Kelly (1993)
	Farming the city	McClintock (2010)
	Land readjustment	Sorensen (2000)
Comprehensive	New urbanism	Larice and Macdonald (2013), Van der Ryn and Calthorpe (2008)
	Agrarian urbanism	Isendahl and Smith (2013)
	Compact city	Burton et al. (2003)
	Evidence-based design	Brown and Corry (2011)

dimension of sustainability (Table 1). It is important to note that certain concepts might place an emphasis on two or more dimensions, and thus by categorizing according to the main focus it becomes clear why not all the problems within cities can be solved through urban planning and design alone. The dimensions (and therefore categories) of sustainability used are: (1) environmental, (2) social, and (3) economic. The discussions of alternative concepts is thus separated into these three categories, with an additional section regarding comprehensive concepts that integrate multiple elements from the main three.

3.2.1 Environmental

New research has highlighted that the best way to build cities might not necessarily be by increasing amounts of hard infrastructure. In particular, ecosystem-based adaptation as well as nature-based solutions provide flexible, cost-effective and broadly applicable alternatives for buffering the impacts of climate change (Maes and Jacobs 2015), and can improve environmental conditions while overcoming negative impacts caused by hard infrastructure (Jones et al. 2012). With this green ecosystem layer, ecosystems-based adaptation can become the major long-term mechanism to ensure the safety of residents (Travers et al. 2013).

It is also important to note how the increase in awareness regarding ecosystem services and its benefits can have a great impact within the various layers of urban planning and design. Alternative urban forms have been proposed around eco-city concepts (Roseland 1997) and landscape urbanism (Waldheim 2006). In general, this type of strategic solution-oriented urban planning, centred around concepts of landscape planning or ecological inputs, can improve the overall sustainability of urban settlements. The problem is that the need for a human scale[2] (Giedion 1967) and traditional mixed use found in cities that have grown organically, were essentially abolished by the monotonous modernist form. Only concepts that directly oppose these notions of a monotonous city, such as the garden city by Howard and Osborn (1965) have been able to achieve coexistence with the environmental dimension. In an attempt to meet the environmental needs imposed by society, various urban planning and design projects across the world are being developed, based on elements or principles of ecological urbanism (Mostafavi and Doherty 2010). In these ecological urbanism projects, landscape architects are first defining which areas can or cannot be built on, based on the natural typography or general landscape characteristics (McHarg and Mumford 1969), rather than having urban planners in charge. This change in the hierarchy of projects, putting landscape architects as the main conceptual decision makers (before urban planners and designers), is a crucial evolution in design, allowing the achievement of a higher level of sustainability. Within this concept, landscape urbanists working as applied "landscape ecologists" must consider sensitivities to ecology that are lost or not considered in traditional urbanization (Mostafavi and Doherty 2010). Charles Waldheim illustrated the need for this paradigm shift: "*The primary complaint has been that landscape architects are taking market-share away from urban designers and planners, but frankly, the urban design and planning realm has been slow to recognize the increasing importance of environmental factors*" (Stein 2010).

Landscape urbanism theory is an important step forward in the urban planning and design professions, as it considers the ecology of the open space, conducting field methodologies that analyze and take into consideration the soil, existing biodiversity, land elevation, and micro-climate during the design phase. Yet, the consideration of environmental concerns has not necessarily meant that social aspects have also been considered, for example in projects where private investors have built paradisiac high standard luxury buildings that fail to address issues of property and social inclusiveness.

3.2.2 Social

There are many mechanisms through which a given project can meet the social dimensions of sustainability. Social inclusion and social control of the urban environment was suggested in the early 1960s by Jane Jacobs in the book *The Death and Life of Great American Cities* (Jacobs 1961). This author proposes several solutions for social problems by observing the city from her window. Not all centered on making physical changes to the city. Through her observations she

[2]A lack of human scale here refers to cities with dimensions either too big or small for humans to use efficiently.

identified social issues that could potentially be solved by changing the morphology of the city. Some of the recommendations included using small building blocks to allow for high pedestrian permeability, changing the types of street use, and having buildings of various ages and states of repair, together with increasing urban density. More recently, other types of solutions have been proposed. Methods include the Machizukuri planning concept that originated from community-based planning (Sorensen 2002). Machizukuri planning does not directly affect the physical shape of the city, and instead allows its residents to co-design the environment according to their needs. A participatory approach, such as developing a collective solution through the Yonmenkaigi system method (YSM) by taking steps in (1) management, (2) information, (3) soft and (4) hard logistics, can make it possible to address short term problems, while simultaneously building solutions to long-term goals (Na et al. 2009). These types of participatory approaches (Machizukuri and Yonmenkaigi) have been proven to be effective. It is argued that they constitute one of the key contributions of sustainability science to assure quality of urban areas (Spangenberg 2011).

3.2.3 Economic

Apart from marketing approaches to urban planning and design there are few planning concepts that are based on economics. As economic competition increases, studies show that cities themselves have become ‘products’, to be sold in an aggressive way in order to attract tourists, residents, commercial activities and investment (Ashworth and Voogd 1990). However, each planning concept or project proposed does have an economic cost or benefit attached to it, and these influence the designs and therefore the total costs of projects.

An example where only the economic cost was considered is Levittown by Levit & Sons between the 1950s and 1960s, which were based on principles of prefabrication and mass-produced low-cost housing (Kelly 1993). The morphological planning and layout of houses was purely functional and based on a road system that allowed for easy automobile access, giving a growing attention to the “centreless cities” and urban sprawl (Frumkin et al. 2004). Such problems could be considered a result of focusing on a single element of functionality, and in such cities farming only appears out of necessity. Although, multiple benefits can arise from urban agriculture, even low income residents find it difficult to increase their self-sufficiency by reclaiming vacant lands in functional cities (McClintock 2010).

Such issues can be partly addressed by methods such as land readjustments (LR), whereby an irregular pattern of agricultural land holdings with different owners is re-arranged into organized building plots and provided basic urban infrastructure such as roads and drains (Sorensen 2000). LR provides the basic infrastructure necessary for new developments, with the intention to make the adjustments as efficient as possible and therefore improve land price (Larsson 1997).

3.2.4 Comprehensive Concepts

In order to enhance the quality of urban planning and design, there have been a range of movements that attempted to simultaneously target two or more of the dimensions of sustainability. Examples of such concepts include New Urbanism (Larice and Macdonald 2013), Agrarian Urbanism (Isendahl and Smith 2013), Compact City (Burton et al. 2003) and others (Fainstein 2000). These concepts are receiving much attention, and new urbanists have been calling for an urban design that includes a variety of building types, mixed uses, intermingling of housing for different income groups, and a strong privileging of the community (Fainstein 2000). Although, the upside of these concepts is yet to prove a success in the long term, and implementations have so far been limited. Critics have commented on how the theory often sounds positive, but the actual implementation and functioning does not work (Newman and Kenworthy 2006).

The need for a change to happen within urban planning and design has been discussed within the professions. Actions of academics and practitioners alike have suggested a need for an academically and practically relevant planning and design profession (Brown and Corry 2011). These authors already recognized the need for evidence and measurability with the characteristics of spatial form (Lynch 1984), though Brown and Corry (2011) clearly state that there is a need for the landscape architecture profession (which is closely linked with the urban planning and design fields) to evolve. It is necessary to follow the trend of a more evidence-based design based on knowledge—generally originating from methodical experiments or experience—as the principal building block (Brown and Corry 2011). For example, virtually every city plan today attempts to create options that reduce traffic and the need for cars. Yet, urban design parameters and concepts for doing this are still not very clear (Newman and Kenworthy 2006).

3.3 *What Does Sustainability Science Focus on?*

Society and its policy makers are calling for a transition towards more sustainable societies. Yet, modern science seems incapable of providing solutions for managing this transition (Dedeurwaerdere 2013). Gibbons et al. (1994) argues that there is a shift from mode-1 to mode-2 science. Mode-1 science is completely academic in nature, monodisciplinary, where scientists themselves are responsible for their own professional performance. Mode-2 science is both inter- and intra-disciplinary, where scientists are part of a heterogeneous network (Gibbons et al. 1994). Such a shift from mode-1 to 2 could potentially help solve the complexities between people and nature. Along with that, the concept of sustainable development has become more prominent since its advent in the 1980s. Evidence-based planning is necessary (Brown and Corry 2011), along with actions that mix together scientific, economic, social and political knowledge (van Kerkhoff and Lebel 2006) for sustainable development to take place. The goals of sustainable development (Kates et al. 2005) are now part of the goals of governments and corporations, and these goals have become core to the tasks of researchers and universities worldwide (Kates 2011).

The emergence of "sustainable development" as a popular concept has revived the discussion about the form that cities should take. It has motivated and provoked scholars and practitioners to seek forms for human settlements that will meet the requirements of sustainability so that built environments can function better than at present. This challenge has brought together scholars, planners, local and international NGOs, civil societies, and governments to propose new concepts and frameworks to restructure urban areas to achieve sustainability (Jabareen 2006).

Sustainability science is usually seen as an emerging and rapidly developing field. It is even seen as the next stage of development in science and generally tries to achieve a holistic view by seeing things through the three dimensions of sustainability, namely (1) social, (2) environmental, and (3) economic (Lozano 2008). Most importantly, the science of sustainability (Spangenberg 2011) also requires quality assurance through extended participation of practitioners and users of the information provided, thereby integrating empirical and anecdotal evidence into a more theoretical frameworks of different disciplines. This approach calls for discussions of and an agreement on priorities based upon the diverse aspects involved (Spangenberg 2011).

Within sustainability science, which is by its nature transdisciplinary, major questions are derived from societal needs, and evidence-based results should assist decision makers to make well-informed decisions (Mauser et al. 2013). Due to the rapid changes within the field, existing institutions have experienced major barriers for the development of sustainability science. Dedeurwaerdere (2013) stressed the need for an in-depth transformation of the modes of organization of scientific research in order to govern the transition to sustainable societies. However, existing analyses of ongoing initiatives have shown that most of the barriers to a major, consolidated effort to introduce the principles of sustainability science to institutions and the society in general will not be removed without far-reaching institutional change. To contribute to the further development of this new field of research, Dedeurwaerdere (2013) suggests that the three core requirements of transformative sustainability science should be applied in an integrated manner. These are (1) an interdisciplinary approach to coupled human/nature systems, (2) an explicit integration of sustainability ethics, and (3) the development of extra-scientific transdisciplinary research collaborations (Dedeurwaerdere 2013).

In particular, sustainability science needs to establish a profound understanding that can be harnessed and used by any society in political or social goals, policies and strategies for tackling a range of sustainability challenges should be formulated, negotiated, implemented, and evaluated. Moreover, in sustainability science, it is expected that interdisciplinary groups of researchers engage in transdisciplinary processes in order to demonstrate how sustainability transitions for society can come about based on a solid problem-solving approach (Jerneck et al. 2011). This problem-solving approach could therefore combine several field methodology tools in order to achieve results beyond morphological planning.

The rapid change in the development of sustainability science has been well documented through an analysis of the citation network (Kajikawa et al. 2007). According to Kajikawa et al. (2007), in the analysis of the citation network urban planning was amongst the 15 main sub-domains active in sustainability science, though the same researcher argues that the term sustainability science is very vague and unclear. There has already been a long debate on how to define sustainability science, and it seems that each discipline has its own sub-definition. However, the trends do show that there has been a larger focus towards global sustainability in recent years (Kajikawa et al. 2007).

The science of sustainability requires not only rethinking the mode of science (either Mode-1 or 2) and its methodologies, but also needs bridging concepts between different disciplines (Spangenberg 2011). This change and use of methodologies has so far been successful in achieving higher quality projects, such as by bridging urban planning and design with environmental, social, and economically related fields. There appears to be further room for bridging with other fields, such as, for example, medicine, technology, and physics. Sustainability science requires re-structuring the fabric of research and practice for all domains of science and engineering, including its respective standard methodologies and institutions. If successfully implemented the value of science would then be expanded for society, public and scientific credibility could be enhanced and a range of newly developed research questions would result (Spangenberg 2011). The field of sustainability science attempts to reach an understanding of the complex and changing interlinkages between natural and human systems in order to transform and develop them in a sustainable manner. For active collaboration with various stakeholders throughout society, "transdisciplinarity" must create an additional critical element within sustainability science (Yarime et al. 2012).

4 Discussion

This discussion aims to identify the next steps that urban planning and design should take to open up the disciplines to new field methodologies that can grasp issues beyond morphological shapes (Moudon 1997). Through this discussion the authors will attempt to find a synergy or potential contributions from sustainability science to urban planning and design. In defence of the modernists it should be said that during their time there was no consideration of sustainable development, as society was focused on achieving strong economic growth. The concept of sustainability science only received large-scale attention since the Brundtland commission in 1987 (Brundtland et al. 1987). The needs of society today have significantly changed, and the world is facing a rapid depletion of resources that is placing pressure on existing concepts of urban planning and design. The authors will thus like to emphasize that urbanization does have a direct influence on social, environmental, and economic aspects. Hence, there is clearly a need for contemporary urban planners and designers to incorporate concepts from sustainability science in order to meet the needs of society.

4.1 What Is Urban Planning and Design Focusing on?

This chapter has shown how urbanization, and therefore the urban planning and design field methodologies, have an immediate effect influencing the different dimensions of sustainability science (Lozano 2008). However, the planning concepts and methodologies that are typically used when designing cities focus either on one dimension only, or fail to holistically address the three dimensions of sustainability science. There is typically a strong focus on environmental issues, e.g. ecological urbanism, eco-city, or social issues, e.g. Yonmenkaigi, Machizukuri, Jane Jacobs, but the gap between them still needs to be closed.

In order to tackle the challenge to achieve sustainable communities, an approach is needed that is not linked to specific disciplines and their respective methodologies. For example, demographic changes affect food security in developing countries, and this is influenced also by rapid urbanization, as it can lead to the loss of fertile agricultural lands in close proximity to cities. Urban design alone, relying in traditional morphological methods, is not suitable to solve such issues and its implications on food security. Alternatives for settlement and food production must be located outside of the city borders, or a mixture of rural and urban may be promoted as a solution. If such schemes do not materialize, arable lands in the periphery of cities will continue to become increasingly urbanized. This was shown in a study conducted by Tan et al. (2005), in which in a megacity such as Beijing about 74 % of new buildings were sitting on former arable lands, and there was a tendency for smaller cities to have even higher percentages (Tan et al. 2005). The transformation process from arable lands to urban areas requires alternative concepts and morphological tools of analysis in order to achieve "soft" solutions, which do not attempt to physically alter the shape of the city with "hard" infrastructure.

Overall, rapid urbanization has not only resulted in a reduction of farmland but also in there being less land available for forests and shrubs in Asian countries (Zhang et al. 2008). Similar trends can be seen in Europe. Belgium, especially the Flemish region, is characterized by a mixture between rural agricultural land and urban settlements. This highly valuable dense mixture offers the possibility to reduce the costs of transportation with agricultural products into the urban (De Zeeuw et al. 2011). However, urban planning policies have called for an "infill development" and a general urban expansion as the way to solve the shortage of housing, instead of attempting costly housing renovation projects. The concept of infill development has been applied since the Flemish Zoning Code (Vlaamse Codex voor Ruimtelijke Ordening) urban planning policy came into force on May 8th, 2009 (Desmet 2010). As a result, local governments have developed residential housing projects while side-lining social and environmental concerns.

In the Japanese context of a rapid aging society and shifting demographics, lower growth trajectories—the slow lane—are increasingly shaping the direction of research. To keep pace with and direct these lower growth trajectories, planners must maximize the potential for slower, sustainable progress. Landscape and urban planning must change direction, if they intend to adjust towards a slower growth (Yokohari and Bolthouse 2011), a recommendation that goes beyond urban morphology and requires thorough scientific research over the effects on the three dimensions of sustainability science (Lozano 2008). In this sense the international community has been urging for the development of improved implementation strategies that can benefit all dimensions of sustainability. The major difference between urban planning and design and sustainability science is that urban planning tends to be focus on methodologies that look at morphology, while sustainability science choses potential methodologies from different disciplines. Essentially, this provides sustainability science practitioners and researchers greater flexibility, potentially improving the overall final quality of projects.

4.2 Where Can Contemporary Urban Planning and Design Derive Inputs From?

In order to improve society, the field methods used by urban planning and designers need to be improved in order to holistically integrate all issues. Geographers, urban planners and designers have already called for new research strategies, with a focus on collaboration by researchers from the natural, social and human sciences and engineering to contribute to a sustainable future (Mauser et al. 2013). Basically, the problem is that urban planning and design methods have fixed analysis methods and application tools. Thus, there is a clear need for a shift in concepts and methods because of the ongoing paradigm of urbanization and due to the fact that urban planning is not ready for the unprecedented negative consequences that can result from it (Mcdonald et al. 2008).

Morphology in urban planning and design clearly plays a major role in shaping the city and community. Physical design studies should not be reduced or abolished, but rather scientific facts should be taken into consideration in order to tackle underlying issues. Within the urban planning and design discipline there are countless examples in which the analysis of urban morphological forms and patterns can be recognized to lead to a better understanding and advancement of the form of the city. However, it is necessary for urban planning and design to become more dynamic, and hopefully a positive evolution in city design can take place by integrating the environmental, economic, and social costs into planning so that a positive evolution in city design can take place.

The end-state design provided by urban planners and designers often does not include the flexibility necessary to mitigate the effects of climate change nor do they consider the increasing intensity of climatological events on urban areas. Adaptation plans to climate change are being implemented more frequently, though they tend to be focused on hard infrastructural methods of adaptation. Depending on each individual case there could be different types of soft methods that are implementable from the perspective of sustainability science.

In order to meet the needs of society and to tackle issues such as resource depletion, environmental pollution, or disasters, new and flexible approaches that address different layers from the discipline of sustainability science can help improve urban planning and design concepts. Sustainability science is ideal to integrate mixed methods to solve various problems before designing specific forms. Examples of these studies include the integration of food security within urban planning concepts (Pothukuchi and Kaufman 1999).

Nevertheless, design aspects will remain important, since consolidating urban areas and improving design is seen to be beneficial not simply from the environmental perspective, but also to improve the 'livability' of urban areas and the provision of services, as well as providing impetus for economic regeneration (Bulkeley and Betsill 2005). Soft measures such as clearly identified evacuation routes have proven their effectiveness during the 2011 Tohoku Earthquake Tsunami to avoid human losses through evacuation (Esteban et al. 2015), yet they have no or little impact on the morphology of the built environment. These are methods that are outside the discipline of urban planners and designers, but have been proven to have significant results in improving resilience. Thus, including them into cities remains a key challenge for researchers working on spatial issues.

5 Conclusion

The authors identified a gap between the scientific knowledge available and the focus and field methods of urban planners and designers, which are typically based on the morphological physical design aspects of planning, while seldom tackling the underlying societal, environmental, and economic issues. When these issues are addressed they tend to be focused on single dimensions and do not integrate all three dimensions of sustainability equally. The authors thus propose that sustainability science can step in and attempt to form a connection, bringing balance to all three sustainability dimensions within the fields of urban planning and design. The ability of sustainability scientists to select methodologies outside its own discipline can lead to new implementations of mixed methodologies and public participation that go beyond the morphological, typological or other physical methodological aspects of urban planning and design (Fig. 1).

Also, the authors conclude that the activities of city planners and designers can inadvertently result in negative effects, due to a focus on city beautification rather than having an evidence-based problem-solving approach. Gentrification is a

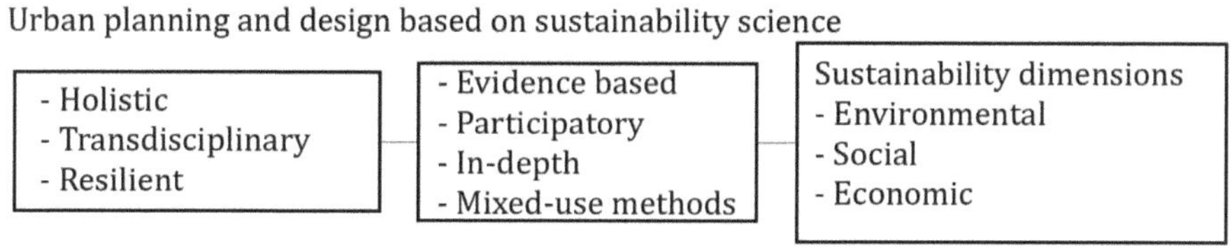

Fig. 1 New concept of sustainable urban planning and designs

highly discussed result of urban interventions, and filling the gap identified within this chapter can potentially lead to a deeper understanding of the misplaced focus of urban planners and designers. In this sense the authors showed that already minor inputs from different disciplines have started revolutionizing the urban planning and design profession.

Understanding the need to tackle the three dimensions of sustainability provides urban designers and planners with an evaluation tool for their own projects. Meeting the needs of each dimension requires designers and planners to reach for solutions outside of their traditional morphological methodologies. Sustainability science provides a framework of mixed methodologies based on a transdisciplinary approach that can help the field truly establish sustainable communities and improve future solutions.

These differences indicate that there is need for a pragmatic approach in which the problems are clearly defined and a solution-oriented science can fill the gap. Pleading both for more accessibility to this knowledge as well as making academic recommendations part of the implementation policies can help achieve sustainable solutions.

It can be concluded that field methodologies used by urban planners and designers need adjustment and require rethinking according to the needs of today. Although urban planners or architects should not necessarily be blamed for all problems in urban areas, it is likely that their field methods (which are mostly focused on objects, volumes, and connections) are too limited for the issues at hand, and therefore new methodologies from sustainability science could be introduced.

Sustainability science could function as the next evolutionary step for urban planners and designers to reach sustainability goals, not just those set out by their own ambitions, but fundamental needs of society such as mitigation strategies, disaster preparedness, food security, and climate change.

References

Albrechts, L. (2004). Strategic (spatial) planning reexamined. *Environment and Planning B: Planning and Design, 31*(5), 743–758.

Al-Soliman, T.M.A. (1988). City beautification, public demand, and environmental priorities in the cities of Saudi Arabia. *Journal of Architectural and Planning Research, 5*(1), 1–13.

Ashworth, G. J., & Voogd, H. (1990). *Selling the city: Marketing approaches in public sector urban planning* (p. 177). Belhaven Press.

Brown, R. D., & Corry, R. C. (2011). Evidence-based landscape architecture: The maturing of a profession. *Landscape and Urban Planning, 100*(4), 327–329.

Brundtland, G., Khalid, M., Agnelli, S., Al-Athel, S., Chidzero, B., & Fadika, L., et al. (1987). Our Common Future ('Brundtland report'). In G. Brundtland, M. Khalid, S. Agnelli, S. Al-Athel, B. Chidzero, L. Fadika, et al. (Eds.). USA: Oxford University Press.

Bulkeley, H., & Betsill, M. (2005). Rethinking sustainable cities: Multilevel governance and the "urban" politics of climate change. *Environmental Politics, 14*(1), 42–63.

Burton, E., Jenks, M., & Williams, K. (2003). *The compact city: A sustainable urban form?* (p. 310). Routledge.

Campbell, S. (1996). Green cities, growing cities, just cities? Urban planning and the contradictions of sustainable development. *Journal of the American Planning Association, 62*(3), 296–312.

Carmona, M. (2010). *Public places, urban spaces: The dimensions of urban design* (p. 330). Routledge.

Changnon, S. A., Kunkel, K. E., & Reinke, B. C. (1996). Impacts and responses to the 1995 heat wave: A call to action. *Bulletin of the American Meteorological Society, 77*, 1497–1506.

de Portzamparc, C. (1995). Des situations plurielles, toujours singulières. L'architecture D'aujourd'hui, 93–94.

De Zeeuw, H., Van Veenhuizen, R., & Dubbeling, M. (2011). The role of urban agriculture in building resilient cities in developing countries. *The Journal of Agricultural Science, 149*(S1), 153–163.

Dedeurwaerdere, T. (2013). Transdisciplinary sustainability science at higher education institutions: Science policy tools for incremental institutional change. *Sustainability (Switzerland), 5*(9), 3783–3801.

Desmet, A. (2010). Vlaamse codex RO—nieuwe decreet plannings-, vergunningen- en handhavingsbeleid. p. 630.

Dodson, J., & Gleeson, B. (2009). *International encyclopedia of human geography* (pp. 77–83). Elsevier.

Esteban, M., Onuki, M., Ikeda, I., & Akiyama, T. (2015). Reconstruction following the 2011 Tohoku Earthquake Tsunami: Case study of Otsuchi Town in Iwate Prefecture, Japan. *Handbook of coastal disaster mitigation for engineers and planners* (pp. 615–631). Elsevier.

Fainstein, S. S. (2000). New directions in planning theory. *Urban Affairs Review, 35*(4), 451–478.

Freestone, R. (2014). Progress in Australian planning history: Traditions, themes and transformations. *Progress in Planning, 91*, 1–29.

Frumkin, H., Frank, L., & Jackson, R. J. (2004). *Urban sprawl and public health: Designing, planning, and building for healthy communities* (p. 368). Island Press.

Gibbons, M., Limoges, C., Nowotny, H., Schwartzman, S., Scott, P., & Trow, M. (1994). *The new production of knowledge: The dynamics of science and research in contemporary societies* (p. 179). Sage.

Giddens, A. (2009). *The politics of climate change* (p. 272). Cambridge, UK: Wiley Online, Library.

Giedion, S. (1967). *Space, time and architecture: the growth of a new tradition* (p. 960). Harvard University Press.

Green, B. N., Johnson, C. D., & Adams, A. (2006). Writing narrative literature reviews for peer-reviewed journals: Secrets of the trade. *Journal of Chiropractic Medicine, 5*(3), 101–117.

Hendrix, J. S. (2006). *Architecture and psychoanalysis: Peter Eisenman and Jacques Lacan* (p. 252). Peter Lang.

Hirt, S., & Luescher, A. (2007). Collaboration between architects and planners in an urban design studio: Potential for interdisciplinary learning. *Journal of Design Research, 6*(4), 422–443.

Holston, J. (1989). *The modernist city: An anthropological critique of Brasília* (p. 383). Chicago and London: The University of Chicago Press Books.

Howard, E., & Osborn, F. J. (1965). Garden cities of tomorrow (Vol. 23, p. 195). MIT Press.

IPCC. (2014). Summary for policymakers. In C. B. Field, V. R. Barros, D. J. Dokken, K. J. Mach, M. D. Mastrandrea, & T. E. Bilir, et al. (Eds.), *Climate change 2014: Impacts, adaptation, and vulnerability. Part A: Global and sectoral aspects. Contribution of working group II to the fifth assessment report of the intergovernmental panel on climate change* (pp. 1–32). Cambridge, United Kingdom, and New York, NY, USA: Cambridge University Press.

Isendahl, C., & Smith, M. E. (2013). Sustainable agrarian urbanism: The low-density cities of the Mayas and Aztecs. *Cities, 31*, 132–143.

Jabareen, Y. R. (2006). Sustainable urban forms: Their typologies, models, and concepts. *Journal of Planning Education and Research, 26*(1), 38–52.

Jacobs, J. (1961). *The death and life of great American cities* (Vol. 71, p. 458). New York.

Jerneck, A., Olsson, L., Ness, B., Anderberg, S., Baier, M., & Clark, E., et al. (2011). Structuring sustainability science. *Sustainability Science, 6*(1), 69–82.

Jiang, Y. (2009). China's water scarcity. *Journal of Environmental Management, 90*(11), 3185–3196.

Jones, H. P., Hole, D. G., & Zavaleta, E. S. (2012). Harnessing nature to help people adapt to climate change. *Nature Climate Change, 2*(7), 504–509.

Kajikawa, Y., Ohno, J., Takeda, Y., Matsushima, K., & Komiyama, H. (2007). Creating an academic landscape of sustainability science: An analysis of the citation network. *Sustainability Science, 2*(2), 221–231.

Kates, R. W. (2011). What kind of a science is sustainability science? *Proceedings of the National Academy of Sciences, 108*(49), 19449–19450.

Kates, W, R., Parris, T. M., & Leiserowitz, A. A. (2005). What is sustainable development? Goals, indicators, values, and practice. *Environment: Science and Policy for Sustainable Development, 47*(3), 8–21.

Kelly, B. M. (1993). *Expanding the American dream: Building and rebuilding Levittown* (p. 287). In W. R. Taylor (Ed.). SUNY Press.

Larice, M., & Macdonald, E. (2013). *The urban design reader* (p. 659). Routledge.

Larsson, G. (1997). Land readjustment: A tool for urban development. *Habitat International, 21*(2), 141–152.

Lozano, R. (2008). Envisioning sustainability three-dimensionally. *Journal of Cleaner Production, 16*(17), 1838–1846.

Lynch, K. (1960). *The image of the city* (Vol. 11, p. 208). MIT press.

Lynch, K. (1984). *Good city form* (p. 528). MIT press.

Madanipour, A. (1999). Why are the design and development of public spaces significant for cities? *Environment and Planning B, 26*, 879–892.

Maes, J., & Jacobs, S. (2015). Nature-based solutions for Europe's sustainable development. *Conservation Letters*, 1–4 (November).

Mahlstein, I., Daniel, J. S., & Solomon, S. (2013). Pace of shifts in climate regions increases with global temperature. *Nature Climate Change, 3*(8), 739–743.

Mauser, W., Klepper, G., Rice, M., Schmalzbauer, B. S., Hackmann, H., & Leemans, R., et al. (2013). Transdisciplinary global change research: The co-creation of knowledge for sustainability. *Current Opinion in Environmental Sustainability, 5*(3–4), 420–431.

McClintock, N. (2010). Why farm the city? Theorizing urban agriculture through a lens of metabolic rift. *Cambridge Journal of Regions, Economy and Society, 3*(2), 191–207.

Mcdonald, R. I., Kareiva, P., & Forman, R. T. T. (2008). The implications of current and future urbanization for global protected areas and biodiversity conservation. *Biological Conservation, 141*(6), 1695–1703.

McHarg, I. L., & Mumford, L. (1969). *Design with nature* (p. 208). American Museum of Natural History New York.

Miller, T. R. (2013). Constructing sustainability science: Emerging perspectives and research trajectories. *Sustainability Science, 8*(2), 279–293.

Mitchell, E. (2009). The new architectural pragmatism: A harvard design magazine reader—William S. Saunders. *Journal of Architectural Education, 62*(3), 92–93.

Mostafavi, M., & Doherty, G. (2010). *Ecological urbanism* (p. 655). Baden, Switzerland: Lars Müller.

Moudon, A. V. (1997). Urban morphology as an emerging interdisiplinary field.pdf. *Urban Morphology, 1*, 3–10.

Moulaert, F., Demuynck, H., & Nussbaumer, J. (2004). Urban renaissance: from physical beautification to social empowerment. *City, 8*(2), 229–235.

Mumford, E. (2002). *The CIAM Discourse on Urbanism, 1928–1960* (p. 375). London, England: Cambridge, Massachusetts.

Na, J., Okada, N., & Fang, L. (2009). A collaborative action development approach to improving community disaster reduction using the Yonmenkaigi system. *Journal of Natural Disaster Science, 30*(2), 57–69.

Newman, P., & Kenworthy, J. (2006). Urban design to reduce automobile dependence. *Opolis: An International Journal of Suburban and Metropolitan Studies, 2*(1), 35–52.
Pothukuchi, K., & Kaufman, L. J. (1999). Placing the food system on the urban agenda: The role of municipal institutions in food systems planning. *Agriculture and Human Values, 16*, 213–224.
Proctor, J. (1998). Ethics in geography: Giving moral form to the geographical imagination. *Area, 30*(1), 8–18.
Roseland, M. (1997). Dimensions of the eco-city. *Cities, 14*(4), 197–202.
Smets, M. (1994). Interview with Manuel de Solà-Morales: The capacity for assessment. Archis, 50–63.
Sorensen, A. (2000). Land readjustment and metropolitan growth: An examination of suburban land development and urban sprawl in the Tokyo metropolitan area. *Progress in Planning, 53*, 217–330.
Sorensen, A. (2002). The making of urban Japan cities and planning from Edo to the twenty-first century. *Nissan Institute/Routledge Japanese studies series* (p. 352). London; New York: Routledge.
Spangenberg, J. H. (2011). Sustainability science: A review, an analysis and some empirical lessons. *Environmental Conservation, 38*(03), 275–287.
Stein, J. (2010). *The rise of landscape urbanism.* Architecture Boston, 38–43.
Tan, M., Li, X., Xie, H., & Lu, C. (2005). Urban land expansion and arable land loss in China—A case study of Beijing–Tianjin–Hebei region. *Land Use Policy, 22*(3), 187–196.
Tibaijuka, A. (2009). A cities and climate change initiative, opening statement. In *Cities and climate change initiative: Launch and conference report* (p. 68). Oslo: UN HABITAT.
Travers, A., Elrick, C., Kay, R., & Vestergaard, O. (2013). *Ecosystem-based adaptation guidance: Moving from principles to practice* (p. 1–97), Working Document. United Nations Environmental Program.
UN-HABITAT. (2009). Planning sustainable cities: Global report on human settlements 2009 (Global Rep) (pp. 1–338). UN-Habitat.
Van der Ryn, S., & Calthorpe, P. (2008). Sustainable communities: a new design synthesis for cities, suburbs and towns (p. 260). Gabriola Island: New Catalyst Books
van Kerkhoff, L., & Lebel, L. (2006). Linking knowledge and action for sustainable development. *Annual Review of Environment and Resources, 31*(1), 445–477.
Waldheim, C. (2006). *The landscape urbanism reader* (p. 288). Chronicle Books.
Watson, V. (2009). "The planned city sweeps the poor away…": Urban planning and 21st century urbanisation. *Progress in Planning, 72*(3), 151–193.
Yarime, M., Trencher, G., Mino, T., Scholz, R. W., Olsson, L., Ness, B., et al. (2012). Establishing sustainability science in higher education institutions: Towards an integration of academic development, institutionalization, and stakeholder collaborations. *Sustainability Science, 7*(SUPPL. 1), 101–113.
Yokohari, M., & Bolthouse, J. (2011). Planning for the slow lane: The need to restore working greenspaces in maturing contexts. *Landscape and Urban Planning, 100*(4), 421–424.
Yokohari, M., Takeuchi, K., Watanabe, T., & Yokota, S. (2008). Beyond greenbelts and zoning: A new planning concept for the environment of asian mega-cities. *Urban Ecology: An International Perspective on the Interaction Between Humans and Nature, 47*, 783–796.
Zhang, H., Ma, W., & Wang, X. (2008). Rapid urbanization and implications for flood risk management in hinterland of the Pearl River Delta, China: The Foshan study. *Sensors, 8*(4), 2223–2239.

A Methodology to Evaluate Sustainability in the Face of Complex Dynamics: Implications for Field Studies in Sustainability Science

Niranji Satanarachchi and Takashi Mino

Abstract Sustainability as a concept has a strong link with the complexity and dynamic patterns of human–natural systems. Evaluating sustainability in human–natural systems requires paying attention to the observation process of these systems to adequately grasp complex dynamics. Failing to do so can result in poor recognition and translation of the sustainability/unsustainability patterns in them. In order to addressing this challenge the present chapter discusses a newly developed methodology to evaluate the sustainability of a human–natural system in a complex dynamic context, which may be useful when conducting sustainability science field exercises. This methodology pays particular attention to the complexities involved in the observation processes, and how awareness of such complexity would support reflexive and iterative understanding-based sustainability evaluations. Finally, the authors will discuss the basis of the evaluation methodology and how it can be applied to field research exercises in Sustainability Science.

Keywords Sustainability evaluation · Human–natural systems · Complexity · A matrix-based methodology · Reflexive and iterative understanding

1 Introduction

In the field of Sustainability Science, '*Sustainability*' has both a conceptual and pragmatic appeal. The conceptual appeal comes from the normative, value laden and contested nature of the concept's definition, which has left room for new interpretations. The pragmatic appeal largely comes from the urgency and problem-oriented nature of the field, which requires the engagement of multiple stakeholders to actively connect problems and solutions. The challenges that the discourse of

N. Satanarachchi (✉) · T. Mino
Graduate Program in Sustainability Science-Global Leadership Initiative,
Graduate School of Frontier Sciences, The University of Tokyo, 5-1-5 Kashiwanoha,
Kashiwa, Chiba 277-8563, Japan
e-mail: nsatanarachchi@gmail.com

M. Esteban et al. (eds.), *Sustainability Science: Field Methods and Exercises*,
DOI 10.1007/978-3-319-32930-7_7

sustainability faces[1] vary among these two ends, yet they also are highly integrated, mainly because sustainability is finally a human interpretation. The close interrelatedness of the conceptual and practical challenges implicitly suggests that these need to go hand in hand to navigate through sustainability problems. Furthermore, preliminary conceptual activities such as framing the problems and identifying multiple systems and their interrelations (or in other words, recognizing complexity) could not be distanced from the activities of practically addressing them. Because of the inherent complex, uncertain, and multifaceted nature of the concept and its problems, it is unavoidable that the frames of observation that researchers and practitioners adopt are limiting and aim at reducing complexity. These limitations and reductions could also negatively affect latter-stage decisions related to implementation, problem analysis or solution. Adopting a frame of observation that is representative enough becomes especially relevant when sustainability is viewed as a process, where the way to understand and change the systems would have dynamic repercussions to its future. Taking these factors into consideration, this chapter will introduce the exploratory sustainability evaluation framework for complex dynamic contexts developed by Satanarachchi and Mino (2014), which may have the capacity to connect the conceptual and empirical ends of sustainability explorations. The authors will discuss the key philosophies that were considered in developing the framework, its steps, and how it could be utilized during the field exercises in sustainability science, particularly for making observations that could lead to a holistic understanding of whether a particular system or community is sustainable or not.

2 Framework to Observe and Evaluate the Sustainability of Human–Natural Systems in a Complex Dynamic Context

2.1 Sustainability of Complex Dynamic Human–Natural Systems

A human–natural system is a unit of understanding the world that implicitly highlights the somewhat separate but interactive importance of humans and the nature

[1]Such as the challenge to observe and understand the complexity and dynamic patterns of human–natural systems and issues regarding their sustainability, or challenges faced during decision-making processes in these systems, that arise largely due to normative standpoints, diverse interests, expertise etc. From the simplest perspective complex dynamics are viewed as patterns in systems that result from the system agents or objects and the interactions among them (derived from definitions by Morin (2008), Miller and Page (2009), Juarrero (2002) and Varela et al. (1974)). For a comprehensive discussion please refer to Satanarachchi (2009, 2015).

that they depend on.[2] Adopting a systemic view to understand humans and their surrounding is not new, and is something that humans do unconsciously all the time. In addition, adopting a systemic view (also known as systems view, or systems perspective) has been recognized as a useful viewpoint in the sustainability discourse, where the interpretations of sustainability need to consider the multitude of interrelated aspects (Clayton and Radcliffe 1996). When undertaking field surveys a practitioner explores both human systems such as social and economic systems, and natural systems such as forest and water ecosystems. These are very much interrelated systems. Not only interrelated, they are also complex dynamic systems (Liu et al. 2007; Ostrom 2007; Holling et al. 2002). Understanding sustainability in human–natural systems involves understanding the complexities and dynamic patterns of these systems. When understanding complex dynamics, one of the essential but often forgotten aspects by mainstream literature is the observation process that allows or inhibits seeing complexity and dynamic patterns. Usually researchers employ a certain frame of observation to observe their surroundings. These frames of observations are conditioned by the researcher's interests, knowledge, beliefs or disciplinary training. Sometimes such ways of seeing the world can help clarify the complexities in the system, but other times they can obscure or hinder a holistic understanding. Sustainability, being a contested concept that encompasses many equally valid but contradicting conditions and directions of systems, needs a holistic approach for evaluations to be made.[3] For an holistic understanding of a system both overarching and specific understanding are equally important. Similarly having both a general and a contextual understanding is important. Sustainability evaluations usually face the challenge of incorporating these equally significant, yet contradicting ends to its assessment processes.

2.2 *Observing Complex Dynamics by Being Sensitive to Complexity*

In order to reach an holistic understanding of the sustainability of a complex dynamic system, it is important to first address the observation process that leads an observer to recognize complexity. Complexity has become a key aspect of human–natural systems, whether they represent the planet as a whole, a country, a region, a town, a village, or a society, as these are not isolated entities but often

[2]Aside from the interaction, which is particularly emphasized in ecology (Liu et al. 2007; Gunderson 2001), the use of the interlinked yet somewhat differentiated term 'human–natural' in this study highlights the fact that sustainability is a human-interpretation that has nature as one of its most important considerations.

[3]In every sustainability-related research or initiative some form of implicit evaluation decision that differentiate sustainability from unsustainability or that differentiates the degree of sustainability is essential.

systems of systems that have dynamic relationships within themselves and their environment.[4] Complexity or complex systems are explored in numerous fields such as mathematics (Engelbrecht 1997; Lorenz 1963), cybernetics (van Dijkum 1997; Gleick and Hilborn 1988), ecology (Holling 2005; Gunderson 2001), social systems (Miller and Page 2009), living system studies (Capra 1996; Goldsmith 1988; Bateson 1979), evolution studies (Hooker 2011; Simon 1991), anthropology (Otto and Bubandt 2010) or philosophy (Morin 2008; Polanyi 1966, 1958); with the interpretations of complexity differing slightly from one field to another.[5] Complexity is often interpreted in terms of parts and wholes (Morin 2008; Heylighen et al. 2006; Ashby 2004; Holland 1998; Cilliers and Spurrett 1999; Simon 1991; Polanyi 1968). For instance it may be interpreted as a characteristic of a system that has a large number of interacting parts that are capable of having organizing relationships (Ashby 2004; Cilliers 2002; Kauffman 1993). Or, it may be seen as the degree to which a system can be differentiated into its parts and can be integrated into wholes (e.g. living systems such as flora and fauna that have nested hierarchical relationships within—individual cells being parts of an organ, organs being parts of a body, and so on). Emergent properties and self-organizing capacity are two key features that can characterize complex systems (Morin 2008; Cilliers 2002; Corning 2002; Goldstein 1999; Holland 1998). Particularly, the self-organizing capacity of complex human–natural systems has attracted considerable attention in scientific inquiries of sustainability (Espinosa et al. 2008; Ostrom 2007; Holling et al. 2002). However, it is also noteworthy that acknowledging factors, such as parts and wholes, and the organizing relationships they generate is unlikely by itself to capture the full implications of complex dynamic sustainability changes in human–natural systems. Sustainability, having a human-interpretation at its core, suggests that the role of the observer (with an observer being defined as an entity who understands and interprets those factors) also largely influences the interpretation. Morin (2008) illustrated three significant stages in observing complexity, using a tapestry as a simile:

(i) In the first stage of complexity, we have simple knowledge that does not explain the properties of the whole. A banal observation that has consequences is not banal; the tapestry is more than the sum of the threads that it is composed of. The whole is more than the sum of its parts.
(ii) In the second stage of complexity, the fact that there is a tapestry means that the qualities of this or that type of thread cannot be fully expressed. The threads are inhibited or virtualized. The whole is therefore less than the sum of its parts.
(iii) The third stage of complexity poses problems relating to our capacity to understand and structure our thoughts. The whole is simultaneously more and less than the sum of its parts.

[4]By dynamic relationships here we generally mean the relationships among the system, system entities and the environment, which can have feedbacks, feed-forwards and emergent properties (Poli 2009; Corning 2002; Goldstein 1999).

[5]For an elaborative review of 'complexity' in these different fields, please refer to Wells (2012).

Morin's (2008) illustrations suggest that in the observation of complexity, observing the whole alone, or observing the parts that build up the whole alone, is not enough. Rather, the observation process must acknowledge that both the parts and the whole have the capacity to inform one another to generate an holistic understanding. The observer's involvement has strong implications for a situation where a system is actually complex but he/she fails to observe it as being complex. Failing to observe complexity could occur in two ways. First the observer could focus on the whole, neglecting the parts (observation supports 'wholistic' understanding).[6] Second he/she could focus on the parts, neglecting the whole (observations support 'partial' understanding). A rigorous analysis of one issue or one aspect of an issue could lead to a comprehensive understanding of that part but still represent a 'partial' understanding of the whole. While there are strengths in both of these ways of looking at a system, in isolation they could lead to an erroneous or incomplete interpretation of a system. Polanyi (1958) gives two scenarios to explain when the observation process plays an illusory role in recognizing complexity, again, with relation to parts and wholes. The first scenario describes the discovery of many prehistoric sites immediately after the airplane was first developed. Although the edges and remnants of these sites had been among human settlements in clear sight for centuries, they were recognized to be cohesive wholes only after their outlines were clearly visible from above. The second scenario is related to his own joint research effort to establish the atomic structure of white tin. During this work Polanyi encountered claims by a different research team, who asserted to have established an entirely different atomic structure to what they had proposed, which later turned out to be an interpretation made along the lines forming an angle of 45° to those along which they had done. This trivial difference in the viewing of the atomic arrangement had rendered it mutually unrecognizable to both parties, simply because of a lack of sufficient understanding of the relationships involved in the atomic arrangement (Polanyi 1958). Polanyi (1958) calls these two scenarios complementary efforts aiming at the elucidation of a comprehensive entity, one that proceeds from the whole towards the particulars, and the other that proceeds from presumed particulars towards grasping the relationship as a whole (Polanyi 1958). His examples suggest that both efforts applied alone would not lead to the elucidation or understanding of a comprehensive entity.[7] This means that in order to have an holistic understanding it is essential that the observation processes that highlight the parts and the ones that highlight the

[6]In this study the term 'holistic' understanding is reserved for interactively generated understanding, through the understanding of the parts that leads to understanding of the whole, and the understanding of the whole that leads to the understanding of its parts. The term 'wholistic understanding' is used for only the understanding that is generated by observing the whole alone. For example, in the case of a bird eye view observation, the understanding could just be confined to the surface of the 'whole'. Having a bird eye view of a field can lead to a 'wholistic' understanding, but not necessarily an holistic understanding.

[7]Polanyi (1958) further names these efforts as two *types of attentions* and discusses their role in the understanding process of a comprehensive entity.

relationship of the parts to the wholes are dynamically integrated,[8] so that they can play a complementary role in generating an holistic understanding. Further, these observations suggest that in order to reach an holistic understanding of the sustainability of a human–natural system it is important to be conscious of how the observations are made.

2.3 A Framework that Supports a Methodology to Evaluate Sustainability

Addressing the need to pay more careful attention to the observation process, the authors developed a framework to evaluate sustainability in a human–natural system, focusing attention on the observation and understanding steps (Satanarachchi and Mino 2014). Its basic structure is illustrated in the Fig. 1.

This framework was developed by considering some of the key requirements of sustainability evaluations, which particularly highlight the researcher's role as an active observing and analyzing agent. The key features of the framework are the focus–system, background layers, sustainability dimensions, sustainability contexts, and sustainability boundaries (Satanarachchi 2015; Satanarachchi and Mino 2014). Focus–system denotes the human–natural system that is being examined to ascertain whether it is sustainable or not, be it a village, town, country or region. Not necessarily confined by administrative or geographical boundaries, it could also represent systems that have functional boundaries (e.g. watersheds), mental boundaries (e.g. mental maps, ideologies) or cultural boundaries (e.g. ethnicities, indigenous communities). The background layers, as the name suggests, denote different backgrounds within which a focused system could be explored to ascertain its degree of sustainability. These backgrounds may include issues (e.g. global warming, nature depletion), systems relationships around those issues (e.g. socio-ecological relationships, socio-economic relationships), key narrations/directions of the system (e.g. economic development, nature conservation, socio-cultural conflicts), key concepts (e.g. growth, stability, resilience) etc. They also could represent different ways of structuring the problems regarding a pressing issue, which can be shaped by different ideological positions, philosophies etc. (e.g. an environmental issue being problematized as a threat to planetary wellbeing, or as a threat to individual wellbeing). When considered over time, they could provide different streams of narrations to interpret the sustainability of the focus–system. Sustainability dimensions represent different angles from which sustainability could be looked at. The dimensions that were proposed for the original

[8]By the term 'dynamically integrated' we mean an integration that occurs as a process. In the two examples by Polanyi this is ensured by a cognitive process supported by various simultaneous cognitive steps such as 'grasping' and 'doing' via sensory organs (Polanyi 1966). In the latter part of the framework, we attempt to create such a dynamic process by directing observations from multiple and conflicting angles (via employing dimensions and background layers).

framework are the (i) sustainability-linked knowledge[9] (ii) sustainability-linked worldviews (iii) resource limitation and availability (iv) wellbeing views (v) policies, rules, regulations, and governing practices (vi) innovations, new creations, and artifacts (Satanarachchi 2015; Satanarachchi and Mino 2014).[10] Collectively, background layers and sustainability dimensions, which are organized and explained under two observational methods (i.e., layer view-based and dimensional view-based methods), enable the observer to decompose the complexity of the system. Further they provide a set of sustainability contexts that can lead to different sustainability boundaries. Sustainability contexts represent detailed interpretations of different observational meta-structures that the observer may come up with after decomposing the complexities of the system (Satanarachchi 2015). A sustainability boundary would demark what is sustainable from what is unsustainable. Therefore, hypothetically sustainability boundaries would enclose a safe operating space for a system.[11] In sustainability discussions it can also represent an evaluation basis that could be derived by looking at one or several of these contextual sustainability interpretations. This process is not a static process. Each sustainability interpretation and evaluation could have the capacity to inform the next interpretation and evaluation, and have a reflexive and iterative relationship. Reflexivity occurs when the attention of the observer simultaneously encompasses both the subject and the object in a way that may challenge the way he/she sees reality (for an elaborative account of reflexivity, please refer to De Cruz et al. 2007; Stirling 2006).[12] Reflexivity can take place in an effective manner when the observer is part of the system being observed. Iteration involves repeating a process continuously, and thus iterative understanding would mean that in each step the previous understanding of the system provides the basis for the next understanding.[13] Together, reflexivity and iteration reflect the relationship between parts and among parts and wholes in a complex system—in this case a complex system of understanding, which is supported by observations. Furthermore, with these understanding processes the observer is transformed from a passive observer to an active observing agent who is sensitive to

[9]By sustainability-linked knowledge we mean the knowledge that is directly linked to sustainability issues, and the systems that are experiencing those issues.

[10]Please kindly refer to Satanarachchi (2015) and Satanarachchi and Mino (2014) for an elaborative account on the rationale and the boundaries of what is explained under each of these dimensions.

[11]A visual representation of the sustainability boundary appears in Satanarachchi (2015) and Satanarachchi and Mino (2014).

[12]Reflexivity has multiple meanings in different fields of studies. For instance it could also denote a characteristic that enables to project the self as an active and creative agent (De Cruz et al. 2007).

[13]Usually the term iteration is used to indicate the act of repeating a process (often given as a function) to reach a certain goal. In this repeating process, the results of one iteration are used as the starting point of the next iteration. Particularly in mathematics, iteration stands for a problem-solving or computational method in which a succession of approximations, each building on the one preceding, is used to achieve a desired degree of accuracy. In an understanding process, the result of one iteration could be seen as a whole that is used as the starting point of the next understanding process.

the complex dynamics of the system. Note that the backgrounds, dimensions, and boundaries in a simple sense provide snapshot views of the system at a certain time. However, what may seem like a snapshot view could encompass information of the system across time, space, and organizing relationships. For instance, sustainability-linked worldviews of a system could easily include time-dependent ideas such as continuity (continuity of tradition that integrates past interpretations of the system present interpretations; continuity of a current state that integrates present interpretations to future interpretations) and survivability (elimination of a future threat that could alter the way one sees present affairs by integrating the future into present interpretations). A background, such as economic development, essentially includes not just the current economic status but how to relate to the past and to the future. Similar patterns could be observed in relation to other dimensions and backgrounds. Such interpretations form multiple partial streams of understanding that the observer makes across space, time, and organizing relationships about the sustainability of the systems. In this way, the framework supports the acquisition of partial streams of understanding of sustainability that may eventually lead to an overall holistic understanding.

3 How the Framework Could Support Field Research in Sustainability Science

Sustainability Science is relatively a new field (Komiyama and Takeuchi 2006; Clark and Dickson 2003; Kates et al. 2001), whose research approaches are often recognized as having a problem-oriented perspective as their starting point (Lang et al. 2012; Swart et al. 2004; Clark and Dickson 2003; Wals and Jickling 2002; Kates et al. 2001). In addition, the importance of a contextual understanding (Robinson 2011; Agrawal 2008) and the capabilities of researchers to gather such contextual understanding (Mino and Hanaki 2013) also are heavily emphasized. Possessing a problem-oriented perspective and focusing on contextual importance means that sustainability research often has at its start field studies. Also, sustainability science practitioners typically start their discussions based on a previous understanding of a pressing sustainability issue. Sustainability issues range from global scale issues, such as global warming, ozone depletion, threats to global ecosystems and biodiversity, ocean degradation, globally significant social issues such as poverty, migration, aging etc., to local scale issues, such as regional resource limitations, local environmental degradation, regional developmental issues, individual and societal issues, local scale technological, knowledge barriers and so on. These issues are not isolated, but they often create complex dynamic patterns by having complementing or contradicting cause-and-effect and feedback relationships. The outward and visible faces of these patterns are what are often recognized as sustainability issues (which relate to certain practices or issues being deemed to be unsustainable, and thus highlighting the need to change in order to move to a sustainable system). Collectively such problems and issues can sometimes lead to the collapse of a given system (referred to as a "system collapse"). Such problems, given the complexities of the interactions and interrelatedness of issues within the system, mean that issues have to be explored across space, time, and organizing relationships for us to obtain an holistic understanding of whether a given system is sustainable or not. Also, it means that contextual complexities need to be untangled to adequately see cause-and-effect and feedback relationships among the issues involved.

As shown in Fig. 1, the framework proposed has basically two distinctive complementary features, namely the layer view-based method and dimensional view-based

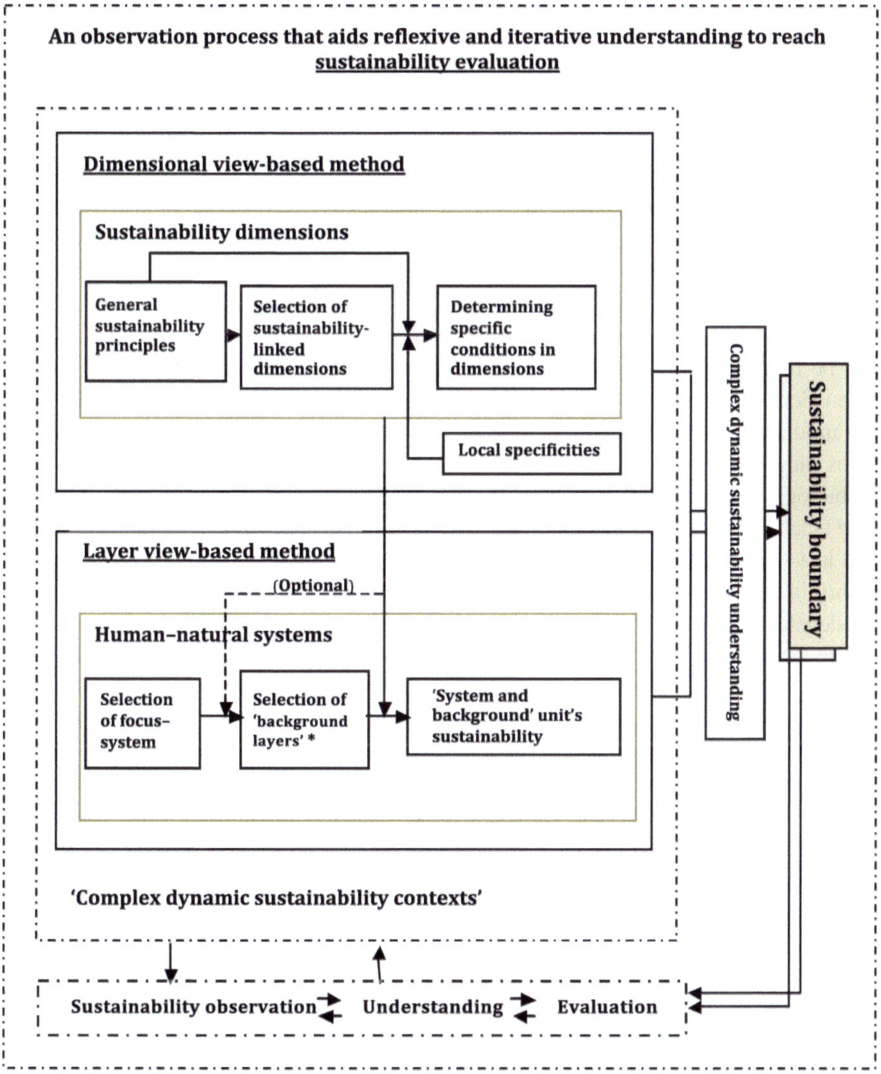

Fig. 1 Framework to observe and evaluate sustainability in complex dynamic contexts (the figure originally appears in Satanarachchi and Mino 2014). *Note* that background layers are selected referring to system relationships and sustainability issues

method. The layer view-based method encourages the observer to select two or more background layers to generally describe the sustainability/unsustainability of the selected focus–system in the field (where the focus–system could be a village, town, watershed, country, region or the globe). These backgrounds could be selected in a way that reflects alternative subsystem interactions (e.g. interaction of nature and individual, economy and individual). Also, backgrounds can be selected based on the issues prevalent in these systems (e.g. nature degradation, poverty). Also, the framework encourages practitioners to think of somewhat conflicting backgrounds in order to better understand the focus–system. We can take some general examples to elaborate on this point. First we could think of a resource-deprived village in the periphery of a preserved natural habitat. For such a village, the lack of economic development and threats to the surrounding natural habitat could be two pressing and also interrelated issues. For geographical reasons, it is likely that given its state of development the village relies on the exploitation of natural resources for its sustenance and economic growth. Additionally, the geography could determine not only its inhabitants' livelihoods but also their social and cultural practices. In such a situation, interpreting the sustainability of the village as a focus–system involves understanding the multiple backgrounds that highlight economic, natural, social, and cultural aspects that would actually provide contradictory backgrounds for sustainability interpretations. An actual example of such a system can be found in the village of Meemure in Sri Lanka (Satanarachchi 2015). This village is located in a valley surrounded by Knuckels forest reserve in the Sri Lanka's central mountain region. The current sustainability problems that the village-forest system faces range from ecological issues in forest (such as excessive logging and threats to its biological and ecological value), to the interlinked socio-economic issues in the village (such as lack of energy and productive land, urban sprawling, population depletion, and loss of historical identity). In order to holistically understand and evaluate sustainability in the village-forest system of Meemure, it was important to look at the historical development of its interconnected and conflicting sustainability issues. In this study the focus–system was selected as the village-forest system, which encompasses its economy, society, and the forest reserve. The key narrations linked to its current sustainability issues—namely socio-economic development and nature conservation—were considered as providing separate but interlinked background layers that could be used for further analysis (Satanarachchi 2015). By observing interlinked backgrounds, it was identified that eliminating one issue could lead to the escalation of the other issue/issues (e.g. forest-conservation efforts had led to resource scarcity and poverty, by denying inhabitants their livelihood), meaning that intervention efforts could actually become drivers of unsustainability change. In such situations sustainability does not mean merely eliminating the issues, but also attempting to identify rapid changes in the system that such interventions could create. In other words, to conduct sustainability evaluations in the face of complex dynamics requires us to consider not only the interactions within the system, but also the interactions between issues across space and time. For that, framing interactions within a focus–system and backgrounds across space and time can be seen as helpful.

Up to the point of selecting the background layers, sustainability remains largely a preliminary understanding by the researcher, often conditioned by immediately visible local sustainability issues and concerns. In order to interpret these sustainability issues, the practitioner would normally rely on some general ideas of sustainability. In addition to such general ideas, the dimensional view-based method suggests that the observer should follow some generally significant principles that reflect multiple facets of sustainability. Earlier we mentioned the dimensions suggested by the framework, which were selected in a way that includes not only the generally significant principles but also somewhat conflicting principles. The users can come up with more dimensions than what was specified, if they see the need to include another significant contextual facet. These dimensions reflect the diversity of angles from which the sustainability of a system could be looked at. Also they play a role in internalizing the external environmental factors into the analysis. For instance, taking into account global discussions and policies on the local issue of forest-cover depletion (including issues relating to knowledge of the global biological significance of the forest reserve in the case described earlier, or knowledge about global warming, such as the carbon capture policies and technology) could provide a different interpretation on the issue from what it would have been if the issue was only considered within the knowledge and political frameworks of the local country/region. Collectively, by including background layers and dimensions, the framework can help to gain multiples contextual interpretations of sustainability (as shown in Fig. 2).

What are the added advantages of adopting a methodology that is sensitive to complex dynamic changes encountered in field research, particularly in the preliminary observation stage? Addressing sustainability in the field—whether it is through the planning of new initiatives, infrastructures, or through specific problem solving—typically requires the involvement of numerous stakeholders from the onset, including planners, policy makers, citizens, experts and analysts. These different stakeholders join the table with different cognitive frames to perceive the system or the problems in the system. Often these frames are hidden and can only be made visible by activities such as dialogue and brainstorming. Soft Systems Methodology (SSM)[14] indicates that a 'methodology' in a transdisciplinary setting, among other factors, is better equipped when having the characteristics of i. visible structure ii. flexibility and iii. explicitness (Scholz and Tietje 2002; Wilson 2001; Checkland 1999) . These characteristics are considered to guide the thinking process and also provide a defensible audit trail (Wilson 2001). The same idea could be applicable to the evaluation of frameworks of sustainability. The proposed framework adopts a structured methodology of differentiating the complexities hidden in general understanding, and then to synthesize the differentiated understanding to once again reach a general, but more holistic understanding. For instance when the framework is applied in the field, it helps to observe a complex reality from multiple angles and to differentiate complexity into a set of

[14]Soft Systems Methodology (SSM) is a methodology that was developed to facilitate preliminary dialogue and brainstorming to tackle problematical (social) situations. It utilizes a systems approach to analyze and solve complex problems, particularly where there are divergent views about the definition of problem (Wilson 2001; Checkland 1999).

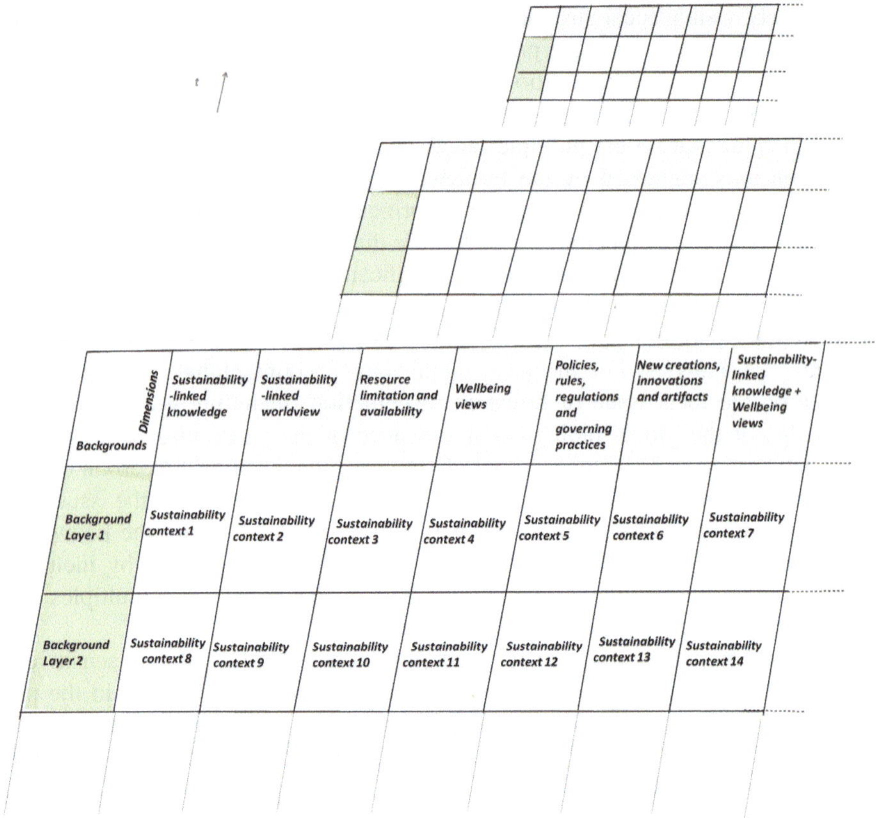

Fig. 2 Matrix to organize sustainability contexts obtained by utilizing the framework in field observation. *Note* Several matrices across time could function as snapshot views of parts that lead to understanding of wholes (sustainability boundaries); for an example of utilizing the matrix in actual empirical analysis please refer to Satanarachchi (2015) and Satanarachchi and Mino (2014)

sustainability contexts (as shown in the matrices in Fig. 2). These detailed sustainability contexts are synthesized back to more general understandings of sustainability boundaries (Satanarachchi 2015; Satanarachchi and Mino 2014). Such a structured methodology would support sustainability science practitioners' efforts to become aware of the thinking processes and the path through which conclusions can be derived. In addition, this process would help not only to untangle the complexities in individual exploration, but also to communicate and to create dialogue in transdisciplinary and interdisciplinary settings.

After a preliminary understanding of a system has been reached by the sustainability scientist, in order to address sustainability problems there is a need to search for further multidisciplinary/interdisciplinary efforts. The dialogues in these latter-stage-efforts are less general in nature than those that supported the preliminary exploration. Because of this it is often the case that preliminary explorations and the subsequent more detailed research can become disconnected. The framework proposed in the present chapter may have the potential to overcome this issue to some extent, as it connects the holistic and specific understanding of sustainability by engaging traceability and reflexivity. Furthermore, the individual sustainability contexts and the boundaries related to each background and the combination of dimension/dimensions could direct the researcher to an in-depth analysis that encourages the exploration of complexities at different levels. This is a desirable situation, as it enables the practitioner to make evaluations based on complex dynamic patterns of the systems while also encouraging appropriate methods and techniques to be employed at different stages of analysis to strengthen and verify their evaluations.

In this way, particularly in field research, the framework functions as a scaffolding to address the complex dynamic aspects of sustainability. In other words, it can help develop the thought processes and structures of a sustainability scientist doing research in the field by encouraging a holistic analysis that relies on key aspects of the process of observation of complexity. While many of its components stay at the abstract level, the framework gains comprehensive meaning after it is applied to a concrete setting. The flexibility that is introduced when selecting the backgrounds and additional dimensions gives room for the explorer/researcher/decision-maker to be actively engaged in the evaluation process, and to reflexively identify the problems that he/she may encounter. In this way, identifying sustainability in the field is no longer treated as merely a complex but static puzzle to be untangled or unfolded, but a complex dynamic process that actively engages the researcher. However, it is also worth noting that the flexibility incorporated into the framework also means that the way in which the framework is utilized would be influenced by the researcher to a large extent. For instance, it leaves room for the researcher to determine the components, such as the background layers and additional sustainability dimensions, which are influenced by the researcher's pre-understanding, views, ability to recognize multiple system relationships etc. Despite this fact, the systemic, interactive, reflexive, and value conscious evaluation process of sustainability, as supported by the framework proposed, may encourage the researcher to also see the multifaceted and uncertain reality in the field.

Foreseeing multifaceted and uncertain aspects and reflexively assessing them is particularly important as sustainability issues could be problematized in different ways, relying on different angles of observation. For a comprehensive understanding of an issue and a subsequent deeper analysis, identifying different narrations from the onset would be helpful. There have so far been two key applications of the framework proposed, related to a socio-ecological system (a village-forest system of Meemure, Sri Lanka, the study appears in Satanarachchi 2015), and a global unsustainability issue (Ozone depletion, see Satanarachchi 2015; Satanarachchi and Mino 2014). In both these studies the framework supported the recognition of conflicting narrations of sustainability, and by doing so enabled the exposure of complexity and dynamic patterns that may have otherwise remained obscured. Furthermore, the sustainability contexts recognized as matrices across time helped to visualize the sustainability boundaries and their change across time (Satanarachchi 2015). In addition to considering past and existing complexities–as was done in these studies–, in the field of sustainability science researchers will often encounter situations where they must envision and design the future of the systems. Envisioning and designing the future indeed becomes crucial in areas such as sustainable urban planning (e.g. envisioning the spaces, interactions of spaces, individuals, networks etc.). In such occasions where field research must engage both the observation of the existing reality and envisioning the future, the systemic and reflexive structuring process supported by the framework proposed in this chapter may help to connect past and present conditions with hypothetical future sustainable or unsustainable scenarios.

4 Conclusion

Sustainability Science has an interdisciplinary and transdisciplinary basis at the heart of its philosophy and research approaches. Therefore, as an academic field it does not possess rigid boundaries, and extensively relies on the frameworks and methodologies developed by other research fields, which are often selected to solve specific problems. They often support generalized overview ideas that overly reduce the complexity in a system, or in contrast, support deep analysis of each part in the system, yet overly isolate the specificities. While both directions have advantages, only partial understanding becomes problematic, especially when sustainability is viewed as a process. In the present chapter the authors introduced a framework designed to overcome these limitations and discussed its potential relevance to field research in Sustainability Science. The key strength of the framework is its ability to help a sustainability practitioner make structured and traceable observations while conducting research in the field. Essentially, the framework has the particular appeal of structuring the messy realities that are usually encountered in field studies, aiding the sustainability scientist to keep a balance between generality and specificity, and enabling the use of specialized methods and techniques while connecting specific measures and evaluations to an overall complex dynamic picture. In addition, it encourages a reflexive understanding process. Reflexivity ensures sensitivity to complexity in an engaged manner and also supports dialogue among stakeholders. In addition to reflexivity, the framework also supports an iterative understanding process, where dynamic observation and understanding processes enable the practitioner to treat sustainability as a multifaceted concept that has great potential for human creativity and design. With these observation and understanding processes, the framework proposed encourages an holistic basis for sustainability evaluations of field exercises in the field of sustainability science.

References

Agrawal, A. (2008). Dismantling the divide between indigenous and scientific knowledge. *Development and Change, 26*, 413–439.
Ashby, W. R. (2004). Classical papers-principles of a self-organizing system. *E-CO Special Double Issue, 6*(1–2), 102–126.
Bateson, G. (1979). *Mind and nature: A necessary unity* (p. 238). New York: Dutton.
Capra, F. (1996). *The web of life: A new scientific understanding of living systems* (1st ed.). NY: Anchor Books.
Checkland, P. (1999). Soft systems methodology: A 30-year retrospective. *Soft systems methodology in action*. Chichester: John Wiley and Sons Ltd.
Cilliers, P. (2002). Why we cannot know complex things completely. *Emergence, 4*(1–2), 77–84.
Cilliers, P., & Spurrett, D. (1999). Complexity and post-modernism: Understanding complex systems. *South African Journal of Philosophy, 18*(2), 258–274.
Clark, W. C., & Dickson, N. M. (2003). Sustainability science: The emerging research program. *Proceedings of the National Academy of Sciences, 100*(14), 8059–8061.
Clayton, A. M., & Radcliffe, N. J. (1996). *Sustainability: A systems approach*. UK: Earthscan.
Corning, P. A. (2002). The re-emergence of "emergence": A venerable concept in search of a theory. *Complexity, 7*(6), 18–30.
De Cruz, H., Gillingham, P., & Melendez, S. (2007). Reflexivity, its meanings and relevance for social work: A critical review of the literature. *British Journal of Social Work, 37*(1), 73–90.
Engelbrecht, J. (1997). *Nonlinear wave dynamics: Complexity and simplicity*. BV: Springer-Science & Business Media.
Espinosa, A., Harnden, R., & Walker, J. (2008). A complexity approach to sustainability–Stafford beer revisited. *European Journal of Operational Research, 187*(2), 636–651.
Gleick, J., & Hilborn, R. C. (1988). Chaos, making a new science. *American Journal of Physics, 56*(11), 1053–1054.
Goldsmith, E. (1988). The way: An ecological world-view. *Ecologist, 18*, 160–185.
Goldstein, J. (1999). Emergence as a construct: History and issues. *Emergence, 11*, 49–72.
Gunderson, L. H. (Ed.). (2001). *Panarchy: Understanding transformations in human and natural systems*. Washington: Island press.
Heylighen, F., Cilliers, P., & Gershenson, C. (2006). Complexity and philosophy. *arXiv preprint cs/0604072*.
Holland, J. H. (1998). *Emergence: From chaos to order*. Reading, MA: Addison-Wesley Helix.
Holling, C. S. (2005). From complex regions to complex worlds. *Minnesota Journal of Law, Science & Technology, 7*, 1.
Holling, C. S., Gunderson, L. H., & Peterson, G. D. (2002). Sustainability and panarchies. *Panarchy: Understanding Transformations in Human and Natural Systems, 63*, 69.
Hooker, C. (2011). Introduction to philosophy of complex systems: A. *Philosophy of Complex Systems, 10*, 3.
Juarrero, A. (2002). Complex dynamical systems and the problems of identity. *Emergence, 4*(1–2), 94–104.
Kates, W. R., et al. (2001). Sustainability science. *Science 292*, 641.
Kauffman, S. A. (1993). *The origins of order: Self-organization and selection in evolution*. Oxford: Oxford University Press.
Komiyama, H., & Takeuchi, K. (2006). Sustainability science: Building a new discipline. *Sustainability Science, 1*(1), 1–6.
Lang, D. J., Wiek, A., Bergmann, M., Stauffacher, M., Martens, P., Moll, P., & Thomas, C. J. (2012). Transdisciplinary research in sustainability science: Practice, principles, and challenges. *Sustainability Science, 7*(1), 25–43.
Liu, J., Dietz, T., Carpenter, S. R., Alberti, M., Folke, C., Moran, E., & Taylor, W. W. (2007). Complexity of coupled human and natural systems. *Science, 317*(5844), 1513–1516.

Lorenz, E. N. (1963). Deterministic nonperiodic flow. *Journal of the Atmospheric Sciences, 20*(2), 130–141.
Miller, J. H., & Page, S. E. (2009). *Complex adaptive systems: An introduction to computational models of social life: an introduction to computational models of social life*. UK: Princeton University Press.
Mino, T., & Hanaki, K. (Eds.). (2013). *Environmental leadership capacity building in higher education: Experience and lesson from Asian program for incubation of sustainability*. Japan: Springer.
Morin, E. (2008). *On complexity*. NJ: Hampton Press.
Ostrom, E. (2007). A general framework for analyzing sustainability of socio-ecological systems. In *Proceedings of the Royal Society B: Biological Sciences* (Vol. 274, p. 1931).
Otto, T., & Bubandt, N. (Eds.). (2010). *Experiments in Holism: Theory and Practice in Contemporary Anthropology*. West Sussex: Blackwell Publishing.
Polanyi, M. (1958). *Personal knowledge*. UK: Routledge and Kegan Paul Ltd.
Polanyi, M. (1966). *The tacit dimension*. UK: Doubleday & Co.
Polanyi, M. (1968). Life's irreducible structure. *Science, 160*(1968), 1308–1312.
Poli, R. (2009). The complexity of anticipation. *Balkan Journal of Philosophy, 1*, 19–29.
Robinson, J. G. (2011). Ethical pluralism, pragmatism, and sustainability in conservation practice. *Biological Conservation, 144*(3), 958–965.
Satanarachchi, N. (2009). *Conceptualizing sustainability dynamics*, Master Thesis, The University of Tokyo, Japan.
Satanarachchi, N. (2015). *Conceptualizing sustainability dynamics: A framework for interface of complex dynamics and sustainability in human–natural systems*, PhD Thesis, The University of Tokyo, Japan.
Satanarachchi, N., & Mino, T. (2014). A framework to observe and evaluate the sustainability of human–natural systems in a complex dynamic context. *SpringerPlus, 3*(1), 1–21.
Scholz, R. W., & Tietje, O. (Eds.). (2002). *Embedded case study methods: Integrating quantitative and qualitative knowledge*. UK: Sage.
Simon, H. A. (1991). *The architecture of complexity* (pp. 457–476). US: Springer.
Stirling, A. (2006). Precaution, foresight and sustainability: Reflection and reflexivity in the governance of science and technology. In J. Voss & R. Kemp (Eds.), *Reflexive governance for sustainable development* (pp. 225–272).
Swart, R. J., Raskin, P., & Robinson, J. (2004). The problem of the future: Sustainability science and scenario analysis. *Global environmental change, 14*(2), 137–146.
van Dijkum, C. (1997). From cybernetics to the science of complexity. *Kybernetes, 26*(6/7), 725–737.
Varela, F. G., Maturana, H. R., & Uribe, R. (1974). Autopoiesis: The organization of living systems, its characterization and a model. *Biosystems, 5*(4), 187–196.
Wals, A. E., & Jickling, B. (2002). "Sustainability" in higher education: From doublethink and newspeak to critical thinking and meaningful learning. *International Journal of Sustainability in Higher Education, 3*(3), 221–232.
Wells, J. (2012). *Complexity and sustainability* (Vol. 26). Canada, NY: Routledge.
Wilson, B. (2001). *Soft systems methodology: Conceptual model building and its contribution*. UK: John Wiley & Sons Ltd.

Sustainability Field Exercises in Rural Areas: Applying the Community Marginalization Framework to Examine Qualitative Changes in Rural Communities

Shogo Kudo

Abstract Rural areas in Japan are facing a critical challenge to their sustainability due to aging and depopulation. These demographic changes have been causing drastic declines in the quality of the living environment, resource management, and social vitality of rural communities. The aims of the present chapter are (i) to discuss the sustainability of rural areas by analyzing the depopulation process of rural communities with respect to their community-functions, and (ii) to address the methodological challenge to examine the sustainability of the target system by applying the community marginalization framework. In this study, the sustainability of a rural area is set as the totality of the self-managing capability of individual communities. To discuss the sustainability of the target system, the methodology employed should be able to examine the changes that occur over time within the system. This chapter showcases how to apply the community marginalization framework in a sustainability science field study, which allows the illustration of qualitative changes in rural communities as they experience population declines over a period of time. The case study was composed of a questionnaire survey to all households in the village. The data collected were analyzed based on five community groups, which were defined by their population sizes. Findings suggested that there is a clear divergence in the process of functional declines among community groups. Especially, communities with less than 40 residents are significantly restricted, regarding their access to basic services, income satisfaction, and the formation of local groups. Moreover, the methodology applied depicted continuing future declining demographic trajectories for the communities with greater than 40 residents, highlighting the severe problems that are being faced by the entire study area.

S. Kudo (✉)
Graduate Program in Sustainability Science-Global Leadership Initiative,
Graduate School of Frontier Sciences, The University of Tokyo,
5-1-5 Kashiwanoha, Kashiwa, Chiba 277-8563, Japan
e-mail: kudo@sustainability.k.u-tokyo.ac.jp

M. Esteban et al. (eds.), *Sustainability Science: Field Methods and Exercises*,
DOI 10.1007/978-3-319-32930-7_8

Keywords Population aging • Depopulation • Community marginalization framework • Rural communities

1 Introduction

1.1 Population Aging as a New Global Trend

Today, world demographic trends are rapidly shifting to a phase of population aging. Some clear evidence of population aging can already be observed and, for example, global life expectancy increased from 47 years in 1950 to 67 years in 2010, and a further increase to 75 years is predicted by 2050 (Crampton 2009; UN 2009). The acceleration of population aging is primarily caused by the declining fertility rate and increasing life expectancy in both developed and developing countries (EC 2009; Lee et al. 2011; Shrestha 2000). While the emergence of population aging itself can be seen as 'a triumph of development', as it implies that a high quality of health, education, and economic well-being has been achieved (Shrestha 2000; UNFPA 2012), aging is expected to have profound impacts on a variety of social dimensions (WEF 2012). Although previous studies have covered diverse topics, they tend to focus on examining the impacts of population aging at the national level, and are primarily concerned with the sustainability of social security systems (Bongaarts 2004; Economic Policy Committee and European Commission 2006; MacKellar 2000; Werblow et al. 2007) and economic decline due to expected labor shortages (Bijak et al. 2005; Thang 2011). In contrast to such macro perspectives, this chapter aims to examine the impacts of population aging from the perspective of rural communities, a micro perspective, through a case study of Kamikoani village in Akita Prefecture, Japan. To achieve this objective, this study will showcase the community marginalization framework, a useful tool for conducting field exercise assessments in sustainability science. Essentially, the community marginalization framework enables the analysis of system quality changes within rural communities. By applying this framework, this study addresses an important methodological challenge when discussing rural sustainability, with a particular emphasis on the time dimension of the target system.

1.2 Population Aging in Japan

Japanese society is rapidly greying, due to the combination of aging and depopulation which is exacerbated by large scale migration of youth to urban areas (Eberstadt 2012). Among all countries in the world, Japan has the highest proportion of 65-plus individuals in its total population (25.1 % in 2015, expected to increase to 38.8 % by 2050). Such rapid population aging is being caused by the country's low fertility rate (1.41 birth per woman in 2012) and long life expectancy (79.94 years for males and 86.41 years for females, as of 2012), resulting in a trend of population decline. The total population of Japan was 128 million at its peak in 2007 (Senno 2013), though it is predicted to become less than 95 million people, or a 25 % decline, by 2050 (MLIT 2011).

Population aging in Japan is happening at an uneven pace and its impacts vary across regions. Particularly, rural areas are experiencing higher proportions of 65-plus individuals than urban areas, and therefore rural communities are considered to be places where the population aging phenomenon can be studied in advance. Such an advanced state of aging societies in rural areas is also the consequence of long-term depopulation due to the excessive out-migration of young individuals from rural to urban areas. This large-scale migration started to take place in the late 1950s to the early 1970s in Japan (Morii 1995). During this period, approximately 1.85 million people migrated to the three main metropolitan areas (Tokyo, Osaka, and Nagoya). According to recent National Census surveys, this out-migration trend is still present, as 38 out of 47 prefectures had negative population growths between 2005 and 2010 (Statistics Bureau of Japan 2011). Even within such prefectures the rates of migration are different, and five prefectures (Aomori, Akita, Iwate, Yamagata, Kochi) had a particularly high depopulation rate during this 5 year period (4–5 % annually). Reflecting such rapid depopulation, the share of age 65-plus individuals in these five prefectures is 26–29 %, 3–6 % higher than the national average (Statistics Bureau of Japan 2011).

1.3 The Community Marginalization Process in the Wake of Rapid Aging in Rural Areas

In recent years, population aging has often been discussed in relation to the functions of local communities in Japan.[1] This perspective has particularly been revived after the 2011 Tohoku Earthquake and Tsunami, as community-based social ties functioned as a prominent means of assistance during the recovery process (Muramatsu and Akiyama 2011). However, the same perspective can be applied for rural communities where various challenges are emerging at the communal level.

In rural communities, residents commonly have neighborhood-based social ties that possess a set of functions for maintaining the local living environment

[1]In this study, "community" is used to refer to as a unit of residence in a given area. Therefore, it is not the intention of this study to discuss the meaning of the term "community" itself, as this term is commonly used as a unit of analysis. In the same manner, "rural" is used only to refer to geographical location of the communities studied (in this sense, this study also does not attempt to study the meaning of the word "rural").

and for organizing local events. These functions, undertaken by local residents, are recognized as "community-functions" (originally "shurakukino" in Japanese) in Japanese rural studies (Kasamatsu 2009; Odagiri 2009, 2011; Ono 1991, 2008; Sakuno 2006). Early studies referred to the community-function as an indicator to assess the state of a given rural community. For example, the management of local resources, seasonal cleanups, and traditional events are considered as activities which represent the quality of community-function at an individual community. A similar concept is also discussed as "community vitality" in the United States. Crandall and Etuk (2008) define community vitality as "the ability of a community to sustain itself into the future as well as provide opportunities for its residents to pursue their own life goals and the ability of residents to experience positive life outcomes" (Crandall and Etuk 2008). The community-function (or community vitality) concept is a useful notion to understand the self-managing capacity of a rural community, and its decline implies the declining vitality of a community.

2 Methodology to Examine the Sustainability of Rural Areas: The Community Marginalization Framework

In this study, the sustainability of rural areas is seen as the totality of the self-managing capability of individual communities. Any qualitative changes in this self-managing capacity over a period of time are considered as a key factor to examine rural sustainability, and thus the methodology to be applied has to be able to analyze the dynamics of this system. Without a methodology including this time dimension, any study on the sustainability of a target system becomes a simple a snapshot or description of the system's current situation. Therefore, it is preferable that a longitudinal study is conducted when the objective of the research is to fully examine qualitative changes of a target system. Yet, it is often the case that "solutions to problems may have to be sought before those problems are sufficiently analyzed or identified" (Komiyama and Takeuchi 2006). In the case of rural Japan, demographic changes, namely aging and depopulation, are taking place at a faster pace than in any other part of the world (Kudo et al. 2015), and hence a novel method that can predict possible trajectories of a target system is needed.

The methodology used in the present study is the community marginalization framework, which has been developed based on the idea of community-functions (Fig. 1), and which enables a researcher to examine changes in rural communities over time. This framework illustrates the declining process of community-functions in a rural community as the community experiences population decline, which is often referred to as community marginalization (Kasamatsu 2009; Odagiri 2009, 2011; Sakuno 2006). The term "marginalization" is derived from Ono (1991, 2008), who coined the term "marginal community",[2] which describes

[2]Originally "genkai shuraku" in Japanese and it means "a community facing its limits".

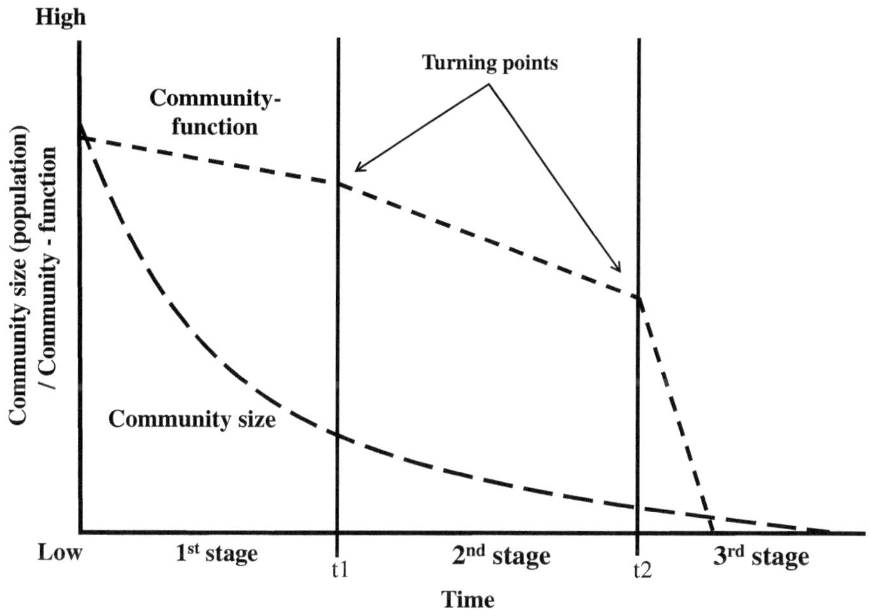

Fig. 1 Conceptual illustration of the marginalization process of rural communities [Modified from Kasamatsu (2009), Odagiri (2009, 2011), and Sakuno (2006)]

the state of a rural community where more than 50 % of residents are over 65 years old. The term describes a situation where significant declines in community-function quality can be observed, so that the residents of a community cannot organize their communal ceremonies by themselves, including for example seasonal gatherings, funerals, or other local events (Ono 2008).

Community marginalization is thus believed to have three stages (Fig. 1). In the first stage, a rural community goes through a drastic population decline, yet the quality of community-functions can be maintained by the remaining community members. In other words, there is still enough population to maintain community-functions. Local activities such as mutual support in production activities, community gatherings, and seasonal traditional events, are maintained, with the content and frequency being preserved in a similar way to what was done by previous generations.

The second stage explains the condition of community in which the size of the population becomes smaller than the required size to maintain the previous degree of community-functions. At this stage some local activities are gradually simplified and integrated with other activities or events. For example, residents agree to do cleanup only once instead of two times per year. It is also common to integrate spring and fall festivals, both of which are used by the community to pray for good harvests, into one festival due to the reduction in cultivated areas.

In a community at the third stage of the marginalization process there are only a few households left. Therefore, the degree of community-functions becomes considerably smaller than in the previous two stages (Fig. 1). At this stage abandonment of local properties such as farmlands, individual houses, and community centers, is often observed. In addition to depopulation, aging affects residents' physical capacity to maintain significant levels of community-functions.

In addition to these three stages of community marginalization, the framework suggests two turning points, which illustrate thresholds between the first and the second stages, and the second and the third stages. As a rural community experiences a decline in its population, it goes through these two tuning points. The framework suggests that the quality of community-functions would significantly differ before and after these turning points. This point indicates possible interventions where rural communities should be re-structured according to the current position of each community in the proposed framework. One of the remaining challenges that are not addressed in literature is how to accurately examine the current position of a rural community within the framework. Nevertheless, the community marginalization framework is considered important as it provides a systemic perspective to understand the transition of a rural community when it is experiencing community marginalization (Kudo and Yarime 2013). In contrast, previous literature focused on each social phenomenon in rural areas such as rural migration (Champion 1989; Hugo and Moren-Alegret 2008; Mitchell 2004), increasing pressure to rural small farmers due to market globalization (Hall 2004; Marsden and Sonnino 2008; Morgan et al. 2010), or general living conditions and well-being of rural elderly (Atterton 2008; Lowe and Speakman 2006; Stockdale 2011). Japanese rural studies have also followed the same trend by examining rural issues such as declining agricultural activities (Ishimaru 2009; Sasaki et al. 2007; Tsukada 1997), declining living conditions in marginal communities (Niinuma 2009; Noguchi et al. 2010; Takegawa 2010), and revitalizing local economy by increasing the interactions with urban residents (Fujita 2005; Tsutsui et al. 2008). The types of challenges that rural communities face today are a mixture of all of these topics, and therefore a more systematic perspective is required. This is particularly the case for the rural areas of Japan, as the combination of aging and depopulation is threatening the sustainability of rural communities.

3 Case Study of Kamikoani Village

To examine community marginalization in rural communities the case study of Kamikoani village (hereafter Kamikoani) in Akita prefecture, Japan, was analyzed. Kamikoani is located in the northern part of Akita prefecture (Fig. 2). Akita prefecture has the highest proportion of age 65-plus residents in Japan, 33.6 % in 2015 against a national average of 25.1 %. In fact, Kamikoani has the highest proportion of over 65's among the prefecture's 25 municipalities, or 50.3 % in 2014. Population aging in Kamikoani has been largely caused by continuous population decline, with the town's 7000 residents in 1960 being reduced to 2524 residents in 2015 (Table 1).

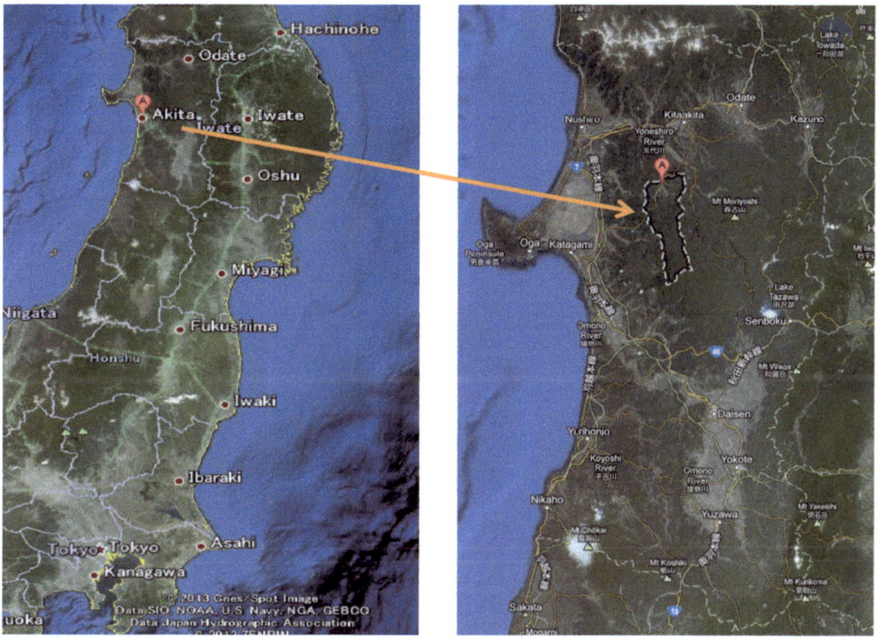

Fig. 2 Location of Akita prefecture and Kamikoani village (*Note* Created from Google map)

Table 1 Demographic trends in Kamikoani from 1955 to 2010

Year	Total population	Population growth rate (%)
1955	6,754	3.4
1960	6,972	3.2
1965	6,550	−6.1
1970	5,242	−20.0
1975	4,708	−10.2
1980	4,352	−7.6
1985	4,116	−5.4
1990	3,746	−9.0
1995	3,553	−5.2
2000	3,369	−5.2
2005	3,107	−7.8
2010	2,727	−12.2
2015	2,524	−7.4

The total area of Kamikoani is 256.82 km^2, which 92.7 % of the land being covered by forests. The primary sector, particularly rice production and forestry, constitutes all major economic activities in the village. However, due to

Table 2 List of communities in Kamikoani and the definitions of the community groups

	Community	Population	Households	Age 65-plus (%)	Community group	Definitions of community group
1	Okitaomote	831	388	47.2	Group 1	Larger than 200 residents
2	Kosawada	371	163	33.2		
3	Hadachi	193	72	45.1	Group 2	Between 100 to 199 residents
4	Shimogotanzawa	179	70	48.6		
5	Oobayashi	151	66	47.7		
6	Fukudate	133	57	50.4		
7	Kamibussha	94	29	33.0	Group 3	Between 60 to 99 residents
8	Doukawa	93	38	26.9		
9	Sugihana	67	28	58.2		
10	Shimobussha	66	25	40.9		
11	Otase	55	21	45.5	Group 4	Between 40 to 59 residents
12	Ooase	52	19	38.5		
13	Nakagotanzawa	48	19	47.9		
14	Daikai	46	18	37.0		
15	Nagashida	45	21	53.3		
16	Minamisawa	36	21	56.4	Group 5	Less than 40 residents
17	Kamigotanzawa	21	11	66.7		
18	Yagisawa	17	8	64.7		
19	Nakamo	6	3	100.0		
20	Fudoura	2	2	100.0		

the national trends of agriculture and forestry decline, the local economy of Kamikoani has also faced a continuous decline.

In order to adopt the community marginalization framework as its methodology, this study divides the communities in Kamikonai into five community groups, which are defined based on their population sizes (Table 2). These community groups (each with a different population) can be shown to depict the "community size" line in Fig. 1, as this line illustrates declining population at the community level. This case study presumes that findings regarding these five community groups will depict different qualities of "community-function" given in Fig. 1. In such a way, the present chapter aims to capture over time changes in rural communities as they go through the community marginalization process.

A household survey was conducted amongst the 1,079 households in Kamikoani.[3] Questionnaire forms were distributed to each of the households, with

[3]There are two elderly welfare facilities in Kamikoani, which have 165 residents in total (each resident in these welfare facilities is counted as one household, and thus there are 165 households between them). Through consultation with the municipal office, it was decided to exclude these households from the survey since they do not join local activities. Thus, although the total number of households in Kamikoani is 1,244, the questionnaire survey only targeted the 1,079 households that do not form part of the welfare facilities, representing a complete survey of the area.

the total number of forms returned being 520 (return rate of 50.0 %). The survey questions were developed regarding five categories: (1) Household profile, (2) Transportation & access, (3) Property management, (4) Economic state, (5) Social relationships. A total of 17 questions were asked, based on a literature review on rural livelihoods (Table 3), which identified a number of topics that are often discussed. Firstly, transportation and access is one of the most discussed topics in early studies. Access to reliable transportation and basic services are seen as key elements for social inclusion in rural settings (Farrington and Farrington 2005; Preston and Rajé 2007). Several other studies have addressed the minimum and affordable size of transportation (Lovett et al. 2002; Smith et al. 2012).

Secondly, the maintenance of both private and community properties is considered as an important topic in rural areas of Japan. Especially, there is high interest regarding the management of abandoned paddy fields (Inaba 2006; Sasaki et al. 2007; Tsukada 1997). Abandonment of private houses and the loss of local culture are also seen as critical issues in rural planning (Shinobe and Miyachi 2012; Ueda 2007; Yusa et al. 2006). In rural planning, a strategic merger of marginalized communities is often discussed (Hayashi 2011; Nishino 2010).

Table 3 Question categories and contents for the questionnaire design

	Question categories	Question number	Contents
1	Household profile	Q1	Members of household
		Q2	Age groups of household members
		Q3	Working members of household
		Q4	Working locations of household
2	Transportation and access	Q5	Main transportation means of household
		Q6	Main means of access to groceries and daily needs
		Q7	Frequency of household's access to groceries and daily needs
3	Property management	Q8	Farm operation by household
		Q9	Removing snow by household
		Q10	Activities or events that household feels lack manpower
4	Economic state	Q11	Working members of household
		Q12	Satisfaction level on current income state of household
5	Social relationships	Q13	Relationships with other members of community
		Q14	Relationships with out-migrated family members of household
		Q15	Frequency of the out-migrated family members' visits
		Q16	Regional activities that household is interested in
		Q17	Willingness to continue to live in the same community

Thirdly, there is a significant amount of public discussions on the issue of local economy revitalization. The economic state of rural areas has been discussed at various levels, from individual to regional. In Europe, migration of retired residents from urban to rural areas is seen as a valuable resource for the revitalization of a local economy, as these people often have established networks from their previous careers (Kalantaridis 2010; Klinthäll 2006; Stockdale 2006).

Lastly, the social relationship among local residents has also been extensively studied in relation to rural livelihoods. Especially, the social capital discourse has gained some popularity. Such studies on social capital examine the relationship between the local environment and the residents' identity formation (Burholt and Naylor 2005; Heenan 2010), and see social capital as the key driver for maintaining rural communities vibrant (McManus et al. 2012).

4 Results

In this section, the results of questionnaire survey regarding four of the question categories are presented. The first category was excluded since the analysis of this chapter is based on the five community groups defined. The results are divided according to community group, which are referred to as Groups 1–5, together with the village total.

4.1 Transportation Means and Access to Basic Services

Regarding the main transportation means of the household, 70–80 % of all respondents answered "Car driven by a household member" (Fig. 3). Aside from that, less than 10 % of households within all community groups use other transportation means. In Groups 1 and 3, about 10 % of households use "Public transportation". In Group 2, 10.5 % of households answered "Walk or bike". In Group 5, 10.5 % of households responded "Car driven by others" as their main means of transportation.

Similar to the results regarding the means of transportation, the access to groceries is largely self-managed within the village. Approximately 80 % of respondents in Groups 1 to 4 said they are "Self-managing" their access to groceries (Fig. 4). Other means are not commonly used, except 11.1 % of Group 4 households, who use "Mobile grocery stores".[4] In the case of Group 5, households who

[4]"Mobile grocery stores" is a type of grocery shopping service which brings products to the community. This service is often offered by a local store and brings those grocery items in a van, so that the customers can shop without traveling to the stores. The difference from a delivery service is that the customers do not need to request a delivery, but can purchase directly while inspecting the product.

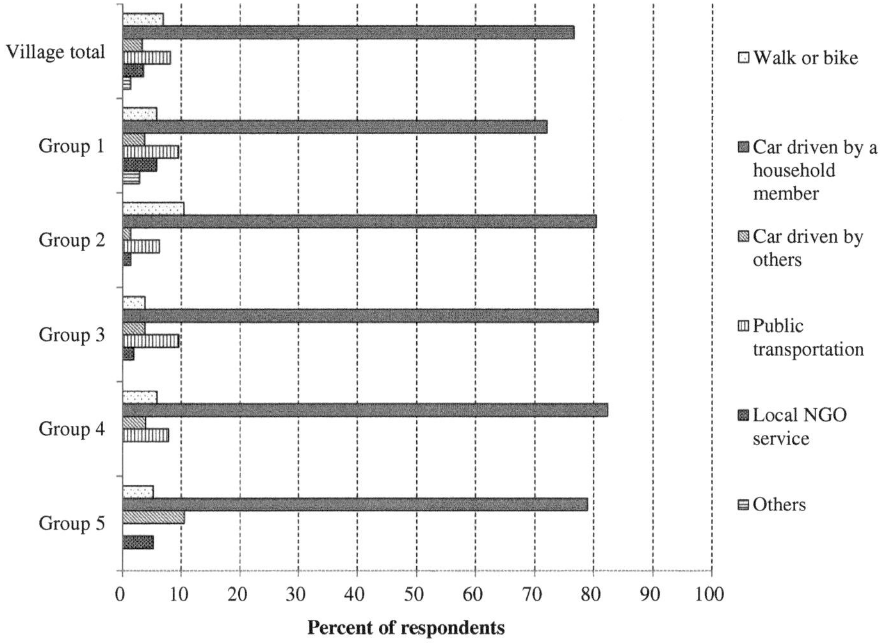

Fig. 3 Main means of transportation of households

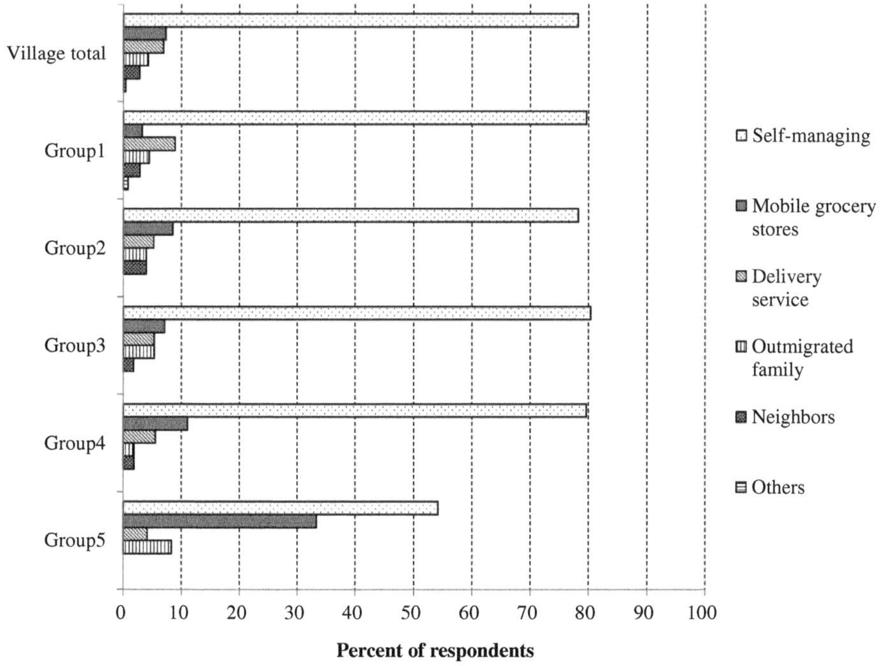

Fig. 4 Main access mean to groceries

are "self-managing" grocery shopping represent 54.2 % of the community, which is about 25 % lower than in other groups. Instead, 33.3 % of households in Group 5 answered that they use "Mobile grocery stores" to purchase daily products. It would appear logical thus that the use of "Delivery services" should also be high in Group 5, however it was highest in Group 1 at 8.9 %.

In terms of the frequency of grocery shopping, a clear difference was found between Groups 1–4 and Group 5. About 45–65 % of households in Groups 1–4 answered that they do grocery shopping "2–3 times per week", while the same response was only 31.8 % in Group 5 (Fig. 5). Instead, 54.5 % of households in Group 5 answered "2–3 times per month".

4.2 Property Management of Households

There are various types of properties in rural areas, including farmlands, community forests, individual houses, and even some intangible resources such as local knowledge and traditions. Since the questionnaire was targeting individual households, the management conditions of their farmlands, individual houses, and seasonal activities were investigated by asking respondents about their current operations.

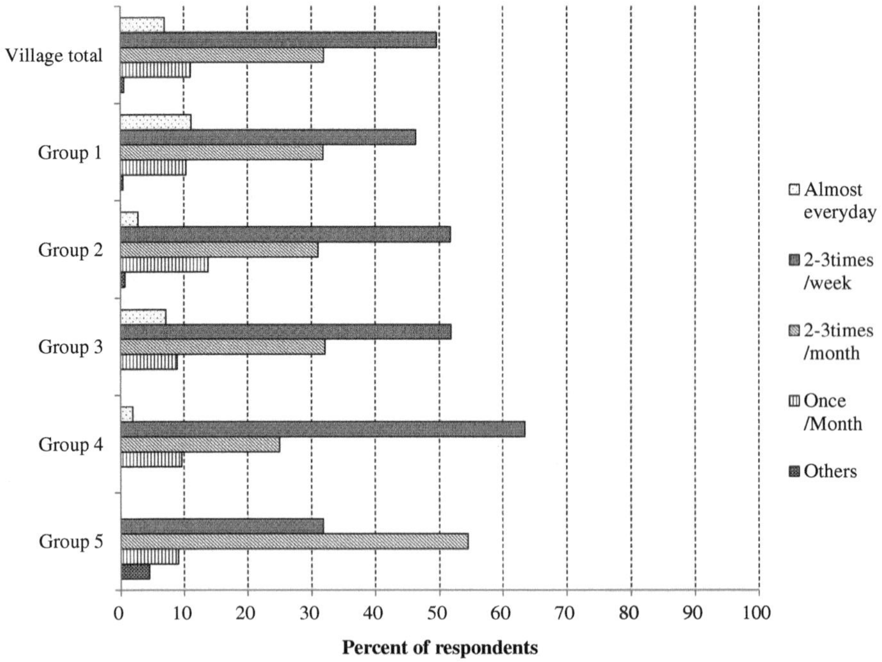

Fig. 5 Frequency of grocery shopping

Regarding the management of farmlands, in Group 1, 45.7 % of households answered that they are "Non-farmers" (Fig. 6). This is largely because communities in Group 1 form the central area of the village, which consists of a higher concentration of built areas than farming areas. Reflecting such characteristics, the most common type of farming was "Self-consumption" in Group 1, at 28.5 %, essentially small scale farming to produce rice and vegetables by households for their own consumption. There are only 6.0 % "Full-time farmer" and 7.1 % "Part-time farmer" households in Group 1.

Compared to Group 1, there are higher proportions of "Full-time farmer" and "Part-time farmer" households in Group 2–4. Especially, the shares of "Part-time farmer" are highest at 25.7 % in Group 3. Also, the share of "Ask other person to cultivate" is higher at 22.6 % in Group 4 and 17.1 % in Group 3. In general, farming operations are fairly well maintained in Groups 2–4, and are especially active in Groups 3 and 4. In the case of Group 5, the type of farming appears different from other groups. The share of "Self-consumption" type of farming is the highest (56.5 %) among all five community groups. In addition, there is no household who answered "Part-time farmer" within this group.

Kamikoani is known for its heavy snow in winter, and therefore removing snow is an important part of maintenance work for individual households. The survey asked how each household is managing this task. Between 70 to 83 % of the households in Groups 1–4 responded that they are "Self-managing". For the case

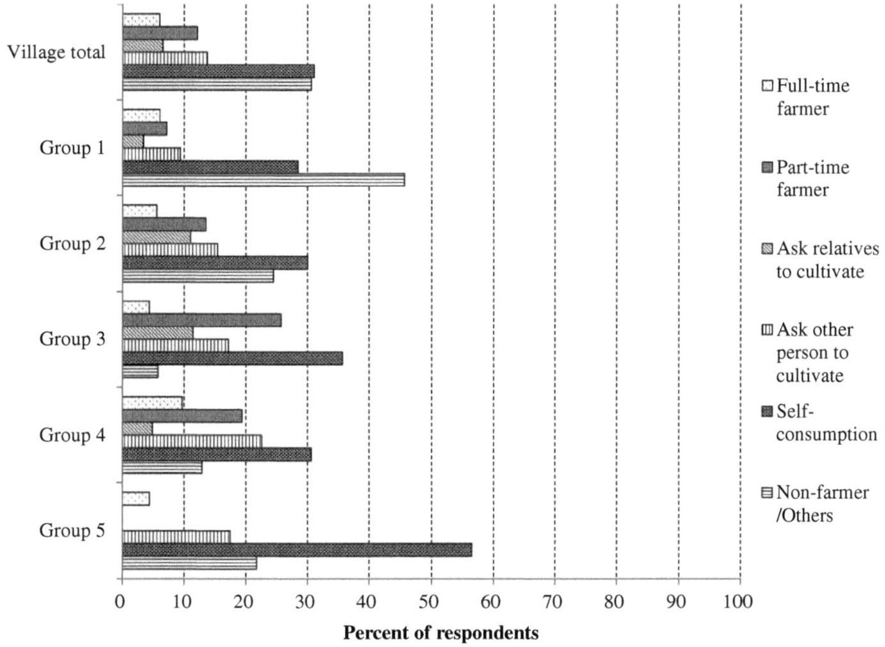

Fig. 6 Farming operation of households

of Group 5, the response of "Self-managing" is slightly lower at 65.2 % (Fig. 7). Instead, 21.7 % of households in Group 5 answered "Ask neighbors" for help, and 13.0 % responded "Ask out-migrated family" members for assistance. Essentially, responses which require the support of others are 5–10 % higher in Group 5 than in the other four community groups.

Another common challenge in property management is the shortage of manpower for local activities. The survey asked the types of activities that residents feel their community does not have enough manpower for. In Group 1–4, between 30 and 60 % of households raised the task of "Removing snow" and between 20–40 % of them responded "Local events" (Fig. 8). The latter answer implies that residents feel they do not have enough people to organize local festivals at the same scale and content as they had in the past. In the case of Group 5, the response appeared different, with 73.9 % of the households mentioning "Removing snow" and only 8.7 % "Local events". Considering the smaller population sizes, this finding implies that local events have already been down-sized significantly in Group 5.

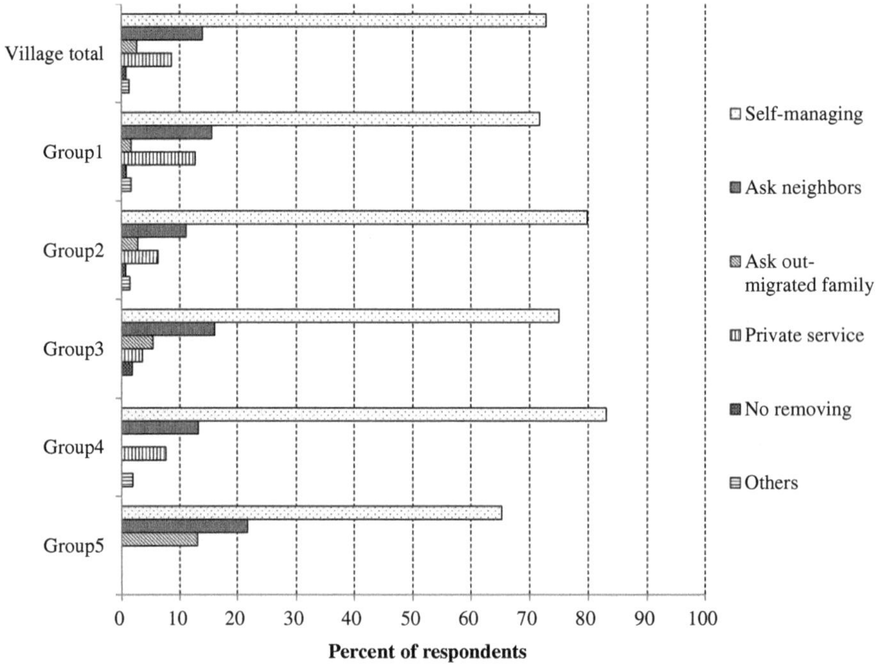

Fig. 7 Removing snow task

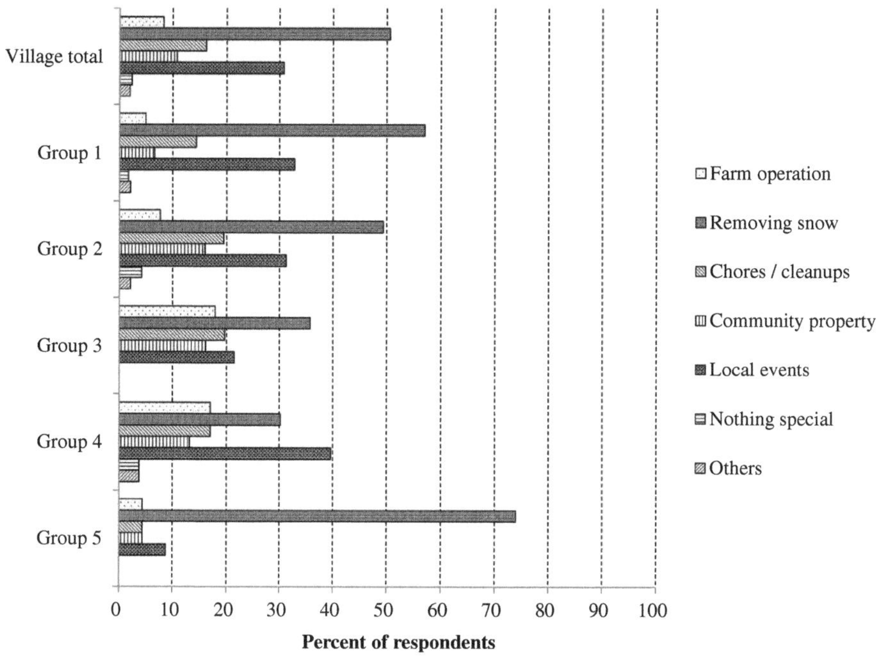

Fig. 8 Activities that residents feel shortage of manpower

4.3 Economic Status: Income Satisfaction Level of Households

To find out about the economic status of households, the questionnaire asked about the income satisfaction level of households. The possible answers provided were "Sufficient", "Sufficient but I want an additional part-time job", "Neither sufficient nor insufficient", "Insufficient so that I need an additional part-time job", "Insufficient so that I need an additional full-time job".

The results show that only 10.9 % of households in the village feel that they have "sufficient" income (Fig. 9), including 10–13 % in Groups 1–3, down to 5.9 % in Group 4, and no household choosing this answer in Group 5. The most common answer was "Neither" at 57.4 % in the village total, which was highest in Group 4 at 62.7 % and lowest in Group 5 at 54.5 %. The total cumulative responses to the answers of "Sufficient", "Sufficient, but I want an additional part-time job", and "Neither" can be seen as the proportion of households that are not expressing negative evaluations on their current income levels. This category of answer consists 70–75 % of Group 1–4 households, while it is only 54.5 % for the case Group 5. Instead, households who expressed their income levels are insufficient so that they need either part-time or full-time jobs are 45.5 % in Group 5, which is about 20–27 % higher than in other community groups. It can thus be concluded that, in general, the majority of households in Group 5 are not satisfied with their current income levels.

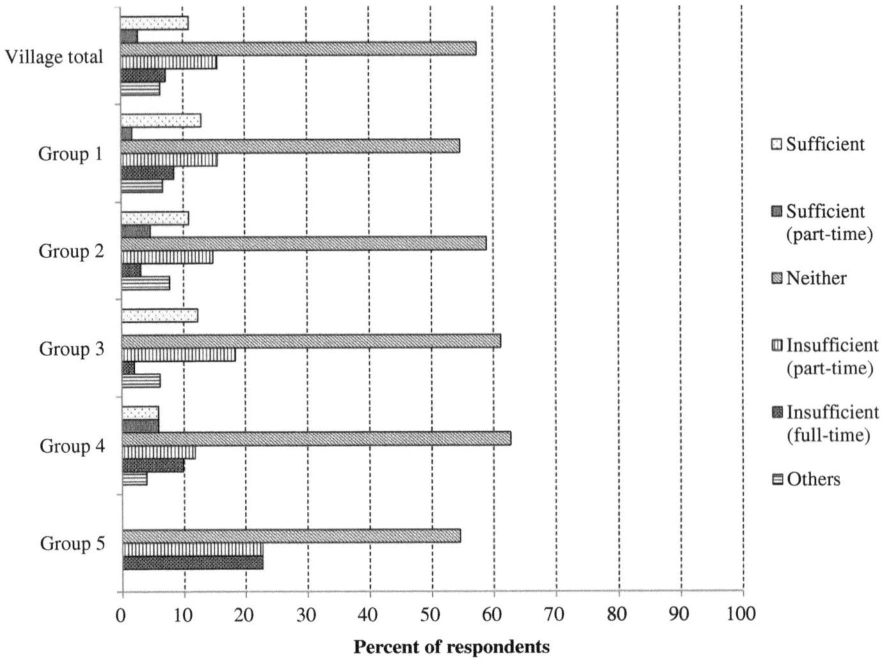

Fig. 9 Income satisfaction

4.4 Social Relationships Among Residents

Social relationships between residents are considered as a key component to enhance local livelihoods in rural communities. The survey examined the household's relationships with others by asking to whom they consult with when they encounter any concerns or problems. In the village total, 41.1 % of households responded that they consult with their "Relatives" (Fig. 10) and 20.7 % with their "Out-migrated children". These two options essentially relate to family members, and consist more than 60 % of the total responses. It is worth noting that in contrast 23.1 % of households answered "None", meaning they do not consult with anyone. The shares of households who said they consult with their "Neighbors", "Town hall", and "Care workers" were all less than 5 %. In Group 1, the most common answer was "None" at 43.5 %, which was the highest among all community groups. In Group 2–5, about 20 % of households chose the "None" answer, while 37–53 % of households in Group 2–5 answered that they consult with "Relatives" (the same answer was only at 21.7 % in Group 1).

In addition to social relationships within each community, the survey asked if households are interested in any particular activities to enhance the local economy and living environment. The possible answers provided in the questionnaire were "Enhancing local market", "Promoting tourism", "Utilize abandoned

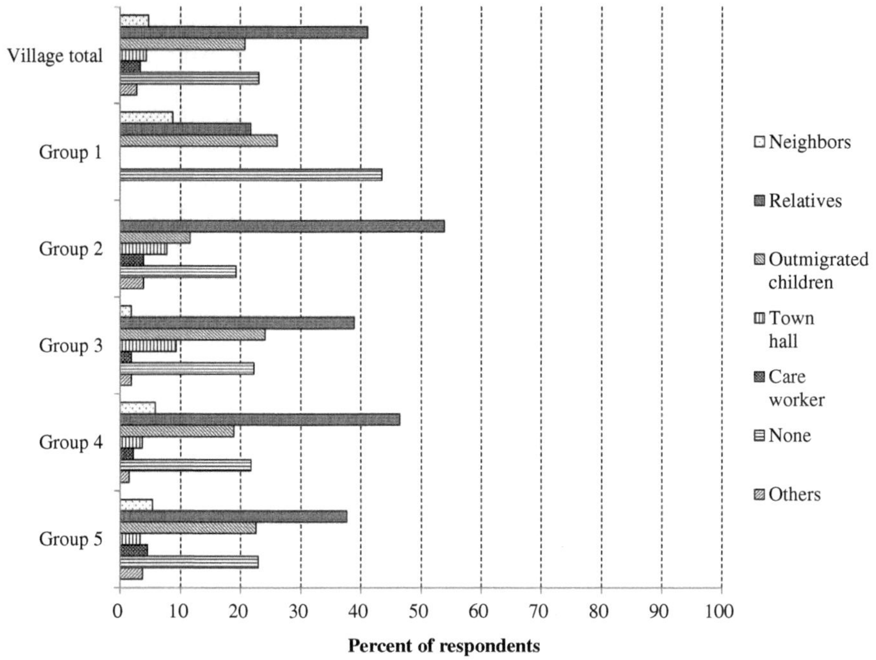

Fig. 10 Person to consult with when having concerns at household

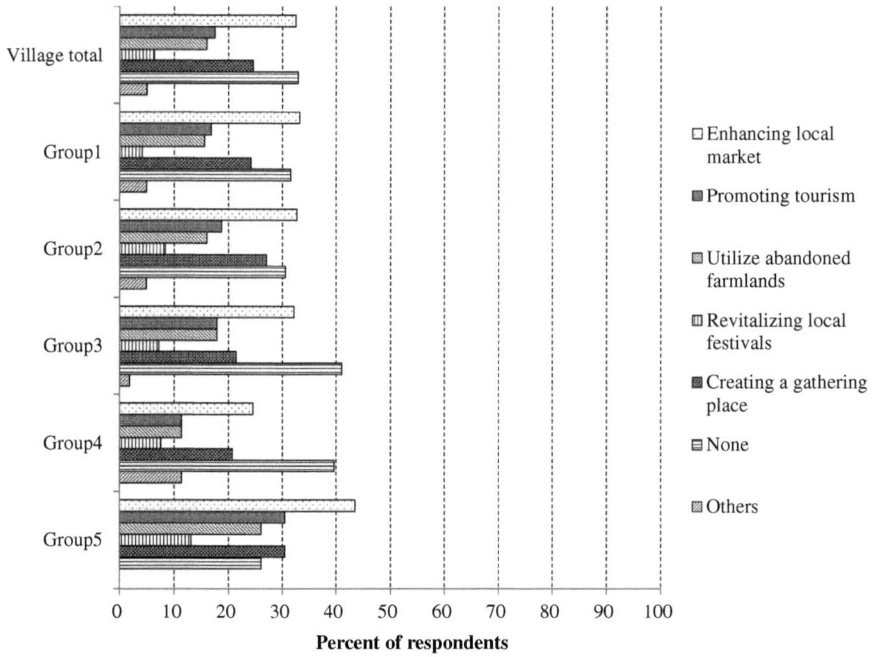

Fig. 11 New activities that households are interested

farmlands", "Revitalizing local festivals", "Creating a gathering place", "None", and "Others". To this question, 25–40 % of the households expressed their interests in "Enhancing local market" (Fig. 11). The second largest response was "Creating a gathering place", in which 20–30 % of villagers were interested. The idea of creating a gathering place aims to increase the interactions among residents beyond individual community boundaries. About 10–18 % of households in Group 1–4 are interested in "Promoting tourism" and "Utilizing abandoned farmlands", while the equivalent responses in Group 5 were 30.4 % and 26.1 %, respectively. Additionally, while many households expressed their interests in new activities, 32.9 % of households in the village answered "None", which implies that there are a significant number of residents who prefer not to start any new activities.

5 Concluding Discussions

5.1 Evidence of the Community Marginalization in Kamikoani

The case study presented in this chapter investigated the current living conditions of households by asking questions regarding five categories across the five different community groups. The results suggest several differences, especially between Groups 1–4 and Group 5, which can be summarized by the following three points. Firstly, the findings about the access to groceries illustrate a higher rate of dependency on outside help by Group 5 households, compared to other community groups. Although the main means of transportation in Group 5 is still a car driven by a household member, a higher rate of household is dependent on mobile grocery stores for purchasing daily items than in other Groups. In terms of the frequency of access to groceries, residents in Group 5 do shopping two to three times per month, while the majority of households in the other four community groups do shopping two to three times per week. Although the findings are limited to transportation and access to daily items, it can be assumed that other issues related to residents' mobility are likely to be present in Group 5, such as the access to care services, medical services, and other public facilities.

Secondly, Group 5 households show unique characteristics in their farming operations, with more than half of those engaged in farming doing so for self-consumption, which does not generate any income. Although around 30–35 % of households in Group 1–4 are also engaged in farming for self-consumption, there are around 7–25 % of part-time farming households in these community groups. In Group 5, there is no part-time farming household. This finding implies a situation that is particular to Group 5, where farming itself is still well-maintained despite its much smaller size.

Thirdly, regarding the "Economic status" of households, the proportion of households who are not satisfied with their current income levels appeared higher in Group 5. Especially, there was no household who responded that they have "Sufficient" income, while about 10 % of households in other community groups are satisfied with their current income levels. Instead, more than 40 % of households in Group 5 feel that the current income level is "Insufficient".

While the current condition of Group 5 was well depicted, differences between the other four community groups could not be so well identified in the present study. This point suggests that a community can maintain its community-functions well until its population size becomes smaller than 40 residents, at least for the case of Kamikoani. Given the clear evidence of community marginalization in Group 5, the size of Group 4, which is 50–60 residents, can be proposed as the pre-stage of Group 5. However, further empirical studies are required to determine the detailed difference between this pre-stage and the marginalized state of a rural community.

5.2 The Current State of Community Groups in the Community Marginalization Framework

The results of present case study showed that the residents of Group 5 are experiencing more difficulties with their living conditions than the other four community groups. Hence, the current state of Group 5 is considered to be in a later stage within the community marginalization framework than the other four community groups. However, at the same time, the Group 5 community is not fully marginalized, as there are various supports provided from outside of communities to maintain the living conditions of residents. Thus, the current condition of Group 5 could be situated right at the end of the second stage of the marginalization process (Fig. 12).

The differences between Group 1 to 3 could not be well identified by this study; instead, their similarities were identified. In these groups, residents expressed greater concerns regarding the shortage of manpower to keep their local events rather than the maintenance of properties. The main means of transportation (car) could still be undertaken by individual households and the access to grocery shopping was more frequent. Based on these findings, and given the fact that they have been experiencing continuous depopulation, these community groups are assumed to be in the middle of the first turning point (Fig. 12). Regarding Group 4, the study predicts that its state can be seen as a pre-state of Group 5, and therefore Group 4 is situated in the middle of the second stage. As the framework shows, the degree of community-functions at this point is still high, as per Groups 1–3. Community-function decline is more difficult to observe at this stage, yet it would be important to highlight that community marginalization could be progressing in communities which are the size of Group 4. As the framework suggests, the degree of community-functions would start to decline gradually during the second stage of the community marginalization. From the perspective of the

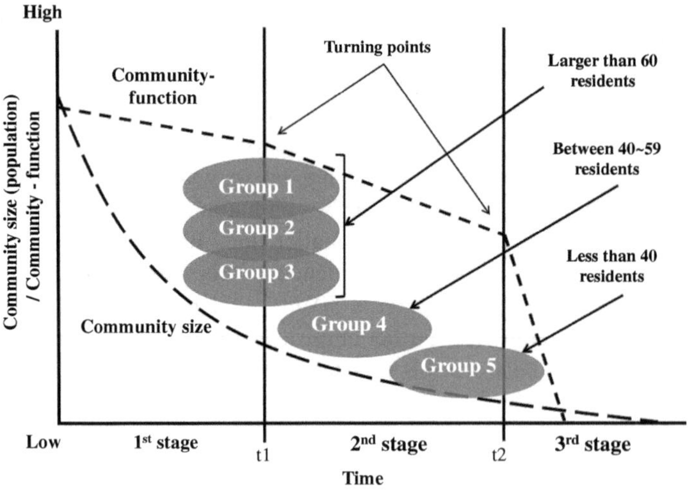

Fig. 12 Current positions of the five community groups in Kamikoani in the community marginalization framework

local residents, this would be observed as a series of downsizing and disappearance of local events and daily activities in their communities.

The community marginalization framework assumes that the livelihoods of households in the third stage become independent from one another, as community-functions nearly disappear at this stage (Fig. 12). However, the case study found that there are always some forms of support provided by relatives or out-migrated children, which can sustain the living conditions of residents. This finding suggests that the current framework of community marginalization limits the perspective to activities within a community, yet in the reality the support provided from the outside should also be taken into consideration. In other words, it would be possible to improve the well-being of residents by enriching their interactions with outside communities. Such intervention approaches needed to be developed further, yet are outside of the scope of the present chapter.

5.3 The Impacts of Population Aging at the Communal Level and the Sustainability of Rural Communities

Based on the results of this study, two types of challenges can be identified regarding the impacts of population aging at the communal level. The first challenge is about property management. Properties in rural areas are not limited to the private properties of households; they include community-owned property such as common forests, community centers, and community cemeteries. These community-owned properties tend to be abandoned as the size of community becomes smaller. The management of private properties gradually changes as owners are no longer capable to maintain them due to aging. This point is proved by the fact that a higher proportion of households in Group 5 maintain their farmlands for "self-consumption" purposes, and that the various supports from outside the community play an important role in maintaining the current quality of living conditions.

The second type of challenge is about local knowledge management. Rural communities often have local traditions which are strongly linked to the management of local resources. Local knowledge is typically managed and transmitted by members of rural communities over generations, though it can slowly fade due to the aging of residents and the marginalization of communities. The current state of rural communities in Japan is rather concerning, and there should be some urgency to record local knowledge before such communities become completely marginalized. While the state of living conditions in marginalizing communities is gaining much attention, including how to provide external support, the management of local knowledge is often given a lower priority. Considering the fact that local knowledge is deeply linked to people's understanding of places, a new approach that includes local knowledge management in addressing marginalizing communities should be developed.

The grand challenge for a rapidly aging (and thus becoming marginalized) rural area is to develop a social system which can preserve some level of community-functions as the population gradually shrinks.. For that, this study identified some of the benchmark changes that residents experience during the community marginalization process. The results also call attention to the need to examine the interactions with the world outside of the communities, as such supports are critical in sustaining the living conditions of the remaining residents. In addition, from the communal perspective, the sustainability of rural communities needs to be discussed not only regarding their physical conditions but also other intangible aspects, such as the social ties among residents and the management of local knowledge. Such a broader perspective is necessary to develop a more inclusive approach for dealing with communities which are becoming marginalized.

References

Atterton, J. (2008). Demographics of the ageing rural population. *Working with Older People, 12*(3), 5. doi:10.1016/j.ypmed.2010.11.009.

Bijak, J., Kupiszewska, D., Kupiszewski, M., & Saczuk, K. (2005). *Impact of international migration and population dynamics and labour force resrouces in Europe* (No. 1). Warsaw. Retrieved from http://www.cefmr.pan.pl/docs/cefmr_wp_2005-01.pdf.

Bongaarts, J. (2004). Population aging and the rising cost of public pensions. *Population and Development Review, 30*(1), 1–23. doi:10.1111/j.1728-4457.2004.00001.x.

Burholt, V., & Naylor, D. (2005). The relationship between rural community type and attachment to place for older people living in North Wales, UK. *European Journal of Ageing, 2*(2), 109–119. doi:10.1007/s10433-005-0028-3.

Champion, T. (1989). *Counterurbanization the changing pace and nature of population deconcentration.* Edward Arnold.

Crampton, A. (2009). Global aging: Emerging challenges. *The Pardee Papers*, (6). Retrieved from http://www.bu.edu/pardee/files/2009/09/pardee_aging-6-global-aging.pdf.

Crandall, M., & Etuk, L. (2008). Whta is community vitality? Retrieved August 17, 2015, from http://oregonexplorer.info/content/what-community-vitality.

Eberstadt, N. (2012). Japan shrinks. *The Woodrow Wilson International Center for Scholars*, 30–37. Retrieved from http://www.wilsonquarterly.com/essays/japan-shrinks.

EC. (2009). *Ageing report : Economic and budgetary projections for the EU-27 member states (2008–2060). European economy*. Luxembourg. doi:10.2765/80301.

Economic Policy Committee and European Commission. (2006). *European economy the impact of ageing on public expenditure: Projectories for the EU25 member states on pensions, health care, long-term care, education and unemployment transfers (2004–2050). Health Care*.

Farrington, J., & Farrington, C. (2005). Rural accessibility, social inclusion and social justice: Towards conceptualisation. *Journal of Transport Geography, 13*(1), 1–12. doi:10.1016/j.jtrangeo.2004.10.002.

Fujita, Y. (2005). コミュニティ・ビジネスが切り拓く地域づくりの可能性 (The potential for regional development through community business). *Tottori University Journal of the Faculty of Regional Sciences, 2*(1), 11–27.

Hall, C. (2004). What does the public want from agriculture and the countryside? A review of evidence and methods. *Journal of Rural Studies, 20*(2), 211–225. doi:10.1016/j.jrurstud.2003.08.004.

Hayashi, N. (2011). 過疎集落からはじまる戦略的な構築と撤退 (Strategic reorganization beginning with a depopulated village). *Journal of Rural Planning Association*, *29*(4), 418–421. Retrieved from https://www.jstage.jst.go.jp/article/arp/29/4/29_4_418/_article.

Heenan, D. (2010). Social capital and older people in farming communities. *Journal of Aging Studies, 24*(1), 40–46. doi:10.1016/j.jaging.2008.09.002.

Hugo, G., & Moren-Alegret, R. (2008). International migration to Non-metropolitan areas of higer income Countries. *Population, Space and Place, 477*, 473–477. doi:10.1002/psp.

Inaba, H. (2006). Measurement analysis of the cultivated acreage and abandoned cultivated lands (in Japanese). *Economic Journal of Chiba University, 20*(4), 79–106.

Ishimaru, N. (2009). 限界集落といわれている集落における耕作放棄地に関する研究 (A study on the situation of abandoned cultivated land in the village, so-called squeezed village—in a case of Soradani District, Aki-Oota-cho, Hiroshima Prefecture). *Architectural Institute of Japan NII-Electronic Library Service, 32*, 4.

Kalantaridis, C. (2010). In-migration, entrepreneurship and rural–urban interdependencies: The case of East Cleveland, North East England. *Journal of Rural Studies, 26*(4), 418–427. doi:10.1016/j.jrurstud.2010.03.001.

Kasamatsu, H. (2009). 小規模高齢化集落の再生にむけて(For regenerations of small scale and aging hamlets). *Bulletin of the Shimane Prefecture Mountainous Region Research Center, 5*, 73–76. Retrieved from http://www.pref.shimane.lg.jp/admin/region/kikan/chusan-kan/syoseki/research/No5_kenkyuhoukoku.data/kenkyuhoukokudai5gou08.pdf.

Klinthäll, M. (2006). Retirement return migration from Sweden. *International Migration, 44*(2), 28.

Komiyama, H., & Takeuchi, K. (2006). Sustainability science: Building a new discipline. *Sustainability Science, 1*(1), 1–6. doi:10.1007/s11625-006-0007-4.

Kudo, S., Mutisya, E., & Nagao, M. (2015). Population aging: An emerging research Agenda for sustainable development. *Social Sciences, 4*(4), 940–966. doi:10.3390/socsci4040940.

Kudo, S., & Yarime, M. (2013). Divergence of the sustaining and marginalizing communities in the process of rural aging: a case study of Yurihonjo-shi, Akita, Japan. *Sustainability Science, 8*(4), 491–513. doi:10.1007/s11625-012-0197-x.

Lee, S.-H., Mason, A., & Park, D. (2011). *Why does population aging matter so much for Asia? Population aging, economic growth, and economic security in Asia*. Manila. Retrieved from http://www.adb.org/publications/why-does-population-aging-matter-so-much-asia-population-aging-economic-growth-and-econ.

Lovett, A., Haynes, R., Sünnenberg, G., & Gale, S. (2002). Car travel time and accessibility by bus to general practitioner services: A study using patient registers and GIS. *Social Science and Medicine, 55*(1), 97–111. Retrieved from http://www.ncbi.nlm.nih.gov/pubmed/12137192.

Lowe, P., & Speakman, L. (2006). The greying countryside. In The aging countryside the growing Older (Ed.), *Population of Rural England* (pp. 9–28). London: Age Concern England.

MacKellar, F. L. (2000). The predicament of population aging: A review essay *. *Population and Development Review, 26*(2), 365–404. doi:10.1111/j.1728-4457.2000.00365.x.

Marsden, T., & Sonnino, R. (2008). Rural development and the regional state: Denying multifunctional agriculture in the UK. *Journal of Rural Studies, 24*(4), 422–431. doi:10.1016/j.jrurstud.2008.04.001.

McManus, P., Walmsley, J., Argent, N., Baum, S., Bourke, L., &, Martin, J., et al. (2012). Rural community and rural resilience: What is important to farmers in keeping their country towns alive? *Journal of Rural Studies, 28*(1), 20–29. doi:10.1016/j.jrurstud.2011.09.003.

Mitchell, C. (2004). Making sense of counterurbanization. *Journal of Rural Studies, 20*(1), 15–34. doi:10.1016/S0743-0167(03)00031-7.

MLIT. (2011). *Long-term Outlook 2011*.

Morgan, S. L., Marsden, T., Miele, M., & Morley, A. (2010). Agricultural multifunctionality and farmers' entrepreneurial skills: A study of Tuscan and Welsh farmers. *Journal of Rural Studies, 26*(2), 116–129. doi:10.1016/j.jrurstud.2009.09.002.

Morii, J. (1995). 高度成長と農山村過疎 *(The Rapid economic growth and the decline of rural areas)* (1st ed.). Kyoto: Bunrikaku.

Muramatsu, N., & Akiyama, H. (2011). Japan: Super-aging society preparing for the future. *The Gerontologist, 51*(4), 425–32. doi:10.1093/geront/gnr067.

Niinuma, S. (2009). Considerations on the maintenance of village functions and the sustainability of residents' daily lives at "Marginal Settlements": The case of settlement M of Hinohara-Mura, Tokyo. *E-Journal GEO, 4*(1), 21–36. Retrieved from http://wwwsoc.nii.ac.jp/ajg/ejgeo/412136Niinuma.pdf.

Nishino, T. (2010). Reorganization of mountain villages of Japan in the early 21st century (in Japanese). *Bulletin of Institute for Research of Regional Economy, 46*(1), 50–66.

Noguchi, K., Sato, S., Kobayashi, Y., Himeno, Y., Siiba, N., & Terada, M. (2010). Classification and characteristics of village by living environment evaluation -living environment and the Sphere in Saiki City, Part1- (in Japanese). *The Japan Association of Economic Geographers*, 3–4. Retrieved from http://ci.nii.ac.jp/els/110008112674.pdf?id=ART0009637691&type=pdf&lang=jp&host=cinii&order_no=&ppv_type=0&lang_sw=&no=1335239391&cp.

Odagiri, T. (2009). 農村再生 *(Regenerating agricultural communities in mountainous regions)*. Tokyo: Iwanami Shoten.

Odagiri, T. (2011). 集落再生 *(Regeneration of rural communities)*. Tokyo: Gyousei.

Ono, A. (1991). 山村の高齢化と限界集落 (Aging in mountain villages and marginal community). *Keizai, 327*, 55–71.

Ono, A. (2008). 限界集落と地域再生 *(Marginal community and regional regeneration)*. Kouchi: Kochi shinbunsha.

Preston, J., & Rajé, F. (2007). Accessibility, mobility and transport-related social exclusion. *Journal of Transport Geography, 15*(3), 151–160. doi:10.1016/j.jtrangeo.2006.05.002.

Sakuno, H. (2006). 中山間地域における地域問題と集落の対応 (The Problems and expectations of regional development in hilly-mountainous region and correspondence of rural settlements). *The Japan Association of Economic Geographers, 52*, 46–64.

Sasaki, H., Koyama, K., & Matsuura, S. (2007). 耕作放棄地の分布と潜在生産力の推定 (Geographical distribution and potential grass productivity of abandoned cultivated land in Japan). *Japanese Society of Grassland Science, 53*(3), 189–194.

Senno, M. (2013). 人口減少社会「元年」は,いつか? "When was the first year of population declining society?" Retrieved January 10, 2014, from http://www.stat.go.jp/info/today/009.htm.

Shinobe, H., & Miyachi, T. (2012). 空き家の解体除去施策の現状と課題-西日本の地方自治体を事例として (The present conditions and problems of the demolition clearance policies of vacant houses). *AIJ Journal of Technology and Design, 18*(39), 709–714.

Shrestha, L. B. (2000). Population aging in developing countries. *Health Affairs, 19*(3), 10. doi:10.1377/hlthaff.19.3.204.

Smith, N., Hirsch, D., & Davis, A. (2012). Accessibility and capability: the minimum transport needs and costs of rural households. *Journal of Transport Geography, 21*, 93–101. doi:10.1016/j.jtrangeo.2012.01.004.

Statistics Bureau of Japan. (2011). *National Census 2010; Executive Summary*. Tokyo. Retrieved from http://www.stat.go.jp/data/kokusei/2010/kihon1/pdf/youyaku.pdf.

Stockdale, A. (2006). The Role of a " Retirement Transition" in the Repopulation of Rural Areas. *Wiley InterScience, 13*, 1–13. doi:10.1002/psp.

Stockdale, A. (2011). A review of demographic ageing in the UK: Opportunities for Rural research. *Wiley Online Library, 221*(October 2009), 204–221. doi:10.1002/psp.

Takegawa, T. (2010). 過疎農山村における高齢者の生活実態と地域福祉の課題-鳥取県日南町における生活実態調査報告- (A study of living conditions of elderly people in depopulating rural districts and challenges of community care: A survey report on living conditions in Nichinan Town, Tottori Prefecture). *Tottori University Journal of the Faculty of Regional Sciences, 7*(1), 2–22.

Thang, L. L. (2011). Population aging, older workers and productivity issues: the case of Singapore. *Journal of Comparative Social Welfare, 27*(1), 17–33. doi:10.1080/17486831.2011.532946.

Tsukada, S. (1997). The revival potential of agriculture and farm villages in Japan (in Japanese). *Association of Economic Geographers, 43*(4), 18–33.

Tsutsui, K., Ebihara, Y., Zushi, N., & Sakuma, Y. (2008). Survey report on a rural community development internship program in Kawamata Town, Fukushima Prefecture (in Japanese). *Tottori University Journal of the Faculty of Regional Sciences, 5*(2), 85–96.

Ueda, K. (2007). About a "Preservation" of traditional dance in a depopulated community: A case study of a "Preservation" of Taiko-Odori in Kajiwata, Kumamoto Prefecture (in Japanese). *Journal of Rural Studies (Japanese), 14*(1), 13–22.

UN. (2009). *World Population Ageing 2009*. New York. Retrieved from http://www.academic-room.com/book/world-population-ageing-2009.

UNFPA. (2012). *Ageing in the twenty-first century: A celebration and a challenge. Ageing in society: An introduction to social gerontology*. New York. Retrieved from http://www.unfpa.org/webdav/site/global/shared/documents/publications/2012/UNFPA-Exec-Summary.pdf.

WEF. (2012). *Global population ageing: Peril or promise?* Geneva. Retrieved from http://www.weforum.org/reports/global-population-ageing-peril-or-promise.

Werblow, A., Felder, S., & Zweifel, P. (2007). Population ageing and health care expenditure: a school of "red herrings"? *Health Economics, 1126*(February), 1109–1126. doi:10.1002/hec.

Yusa, T., Goto, K., Kurauchi, D., & Murakami, K. (2006). 中山間地域における空き家およびその管理の実態に関する研究-山梨県早川町を事例として- (Vacant houses and their management in hilly rural areas—a case study in Hayakawa Town, Yamanashi Prefecture). *Journal of Architecture and Planning2*, (601), 111–118.

Participatory Mapping and Problem Ranking Methodology in the Research of Sustainable Communities—Workshop with Indigenous People Under Community-Based Forest Management Program in the Philippines

Marcin Pawel Jarzebski

Abstract This chapter aims to discuss a methodology for envisioning the sustainable development strategies that can be adopted by indigenous communities, through the use of Participatory Mapping and Problem Ranking. Such methods can facilitate the discovery of the local context in sustainability research, based on a review of the present situation in a given community, and an analysis of their recent past. Essentially, the local conditions inherent to a community need to be systematically analyzed to achieve a higher level of participatory development. The main purpose of this study is to explain how to implement these two methods. To showcase such methodologies the chapter will use the example of community forestry contracts concluded with indigenous groups in the Philippines, and review whether the Community-Based Forest Management policy implemented really helped achieve better sustainable development in the local context. The results of this Participatory Mapping and Problem Ranking exercise revealed that the community forestry program was not well suited to local conditions in which there was lack of sufficient understanding of the land ownership and local problems upon the Community-Based Forest Management project's implementations, and thus a revision of the policy and implementation guidelines is needed so that its primary goal of promoting sustainable development could be achieved.

Keywords Participatory mapping · Problem ranking · Sustainable community development · Indigenous people

M.P. Jarzebski (✉)
Graduate Program in Sustainability Science-Global Leadership Initiative,
Graduate School of Frontier Sciences, The University of Tokyo,
5-1-5 Kashiwanoha, Kashiwa City 277-8563, Japan
e-mail: marcin.jarzebski@yahoo.com; marcin.p.jarzebski@gmail.com

M. Esteban et al. (eds.), *Sustainability Science: Field Methods and Exercises*,
DOI 10.1007/978-3-319-32930-7_9

1 Introduction

1.1 Development and Self-determination Issues of Indigenous Communities

Discussing the sustainability of communities is a challenge both from the ethical and practical perspectives, especially whenever the subjects of such debate are indigenous people. Indigenous people are at the crossroad of traditional and modern development concepts (Viano 2015; Ziai 2015), existing in the legal pluralism, in which both the legal systems of modern states and indigenous law govern their communities (Napoleon 2013). Legal pluralism does not only concern social structures, but also challenges natural resource governance rights and land ownership (Bavinck and Jyotishi 2015). Local resources are still the source of livelihood of many people, and thus rights to access and use them are crucial. Based on the nature of their use of local resources there are several possible pathways a community can take: namely localization, globalization, or glocalization (Wilson 2012; Inoue et al. 2015). Localization, or in other words a 'resistance' strategy, depicts maintaining autonomy and resisting to adapt to globalization; globalization, or the 'adjustment strategy', represents the communities that are willing to benefit from the global economy and protect local resources. Glocalization, which is a compromise between the first two strategies, enables collaboration of stakeholders and collective management of resources (Inoue et al. 2015).

During postcolonial decentralization movements the partial or ultimate rights of forest resources were returned to communities in some countries, in the form of community forestry or the recognition of indigenous people's rights. Community forestry can be broadly defined as the collective management of forests and rural development through the common use of local forest resources, together with the development of alternative income generation means and/or the development of infrastructure (Food and Agriculture Organization 1992; Inoue 2003). However, in certain situation the concept of "community-based" is actually unknown to the indigenous people, and thus Decentralizing power over the forest to these groups may not be a solution (Howitt and Suchet-Pearson 2006; Veland et al. 2013). Recognition of the rights of indigenous people is slowly progressing, and has gained some momentum through the Declaration on the Rights of Indigenous Peoples by the United Nations in 2007. The declaration, although not legally binding in itself, is the legal foundation of indigenous peoples' rights to recover ancestral sovereignty and self-determination (Tobin 2014).

1.2 Participatory Appraisal of Sustainability

The methods of envisioning future sustainable development, such as scenario-based participatory backcasting, are already well established. Backcasting is a way of understanding a possible, desirable future and afterward designing policies and programs to achieve that goal (Vergragt and Quist 2011; Robinson et al. 2011). However, unlike backcasting, which refers to a point in the future to the present, the current analysis of the problems in a community with regards to a reference point in the past still needs to be more systematically tackled, in order for it to provide meaningful answers to sustainability science research. The local conditions that a community possesses are essential for participatory development to be achieved (Platteau and Abraham 2002). Two methods that could determine local conditions, recent changes in the community and obstacles to further development are known as Participatory Mapping and Problem Ranking (Rietbergen-McCracken and Narayan-Parker 1998).

1.2.1 Mapping Socio-cultural and Geographical Spaces

Participatory Mapping is one of the tools developed to integrate the knowledge and ideas of rural people into the planning and management of development projects (Narayanasamy 2009). Participatory Mapping enables a local community to explain various aspects of an area through drawing maps, and facilitates the understanding of a community's situation and conditions, providing not only the physical but also the socio-economic background to it (Narayanasamy 2009). It shows the different types of natural resources and local environments, with the mapping process helping to discuss opportunities and constraints, generating a variety of data in a short time (Mukherjee 2002; Rietbergen-McCracken and Narayan-Parker 1998). More details on the process of Participatory Mapping are discussed in the Sect. 2.3.

Participatory Mapping has often followed post-colonial movements in which maps were associated with the territories of indigenous people, who challenged central government to transform cultural landscapes[1] into areas self-governed by the indigenous community, e.g. in Canada (Bryan 2015). On the other hand, in the contrary case (i.e. when the central government attempts to delineate the territories of indigenous people), it has been argued that the application of the Participatory Mapping is not able to truly reflect community borders (Bryan 2011). Community borders, e.g. in form of zones, simply can not be defined in the neo-liberal sense such as demarcated aministrative borders and doing so may lead to a rise in conflicts between stakeholders (Reyes-García et al. 2012). In the current research Participatory Mapping is applied to the mapping of natural resources, including their actual location and present and future use scenarios (Levine and Feinholz 2015).

1.2.2 Problem-Based Analysis

Problem-based learning in the process of participatory analysis for sustainable development is considered as the foundation for the identification of sustainability

[1]Cultural landscape is a "combined works of nature and of man" (UNESCO 2008).

indicators and as a path leading to community empowerment (Fraser et al. 2006). The Problem Ranking method is a simple and direct way to identify in a more systematic manner problems and impacts that could be overseen during interviews and discussions. It is a technique that involves eliciting the most important problems that local people face, which relies on them self-reporting such problems (Rietbergen-McCracken and Narayan-Parker 1998). The process of Problem Ranking is discussed in the Sect. 2.4.

1.2.3 Combining the Participatory Mapping and the Problem Ranking

The two participatory appraisal methods conducted together can be beneficial for perusing sustainable development goals and examining their applicability to the specific community. Combining both methods can be mutually beneficial. The Problem Ranking is able to identify issues that may be interrelated with spatial and socio-cultural spaces mapped through the Participatory Mapping. These two methods can elucidate the current situation of a specific community, uncovering potential development based on directions that the communities undertake, and the pace of changes taking place.

1.3 Delineation and Granting Access to the Forest—The Philippines Experience

The Philippines is one of the leading countries in the area of community forestry, in light of its progressive policy and the length of its experience. The Philippine community forestry policy includes complete forest devolution—a process of passing rights and decision power to local users in the post-colonial era (Inoue 2003; Pulhin 2003; Shivakoti et al. 2015). In 1995 a new scheme of Community-Based Forest Management (CBFM) was enacted. This scheme took the form of a land tenure system, including forest and farms in the forest's vicinity, with the certificates of CBFM being issued by the Department of Environment and Natural Resources (DENR, See Pulhin et al. 2015). In fact, indigenous people's claims on land, including forest, have been checked by DENR since 1993, resulting in the issuance of Certificate of Ancestral Domain Claims (CADC) or Certificate of Ancestral Lands Claims (CALC) (Magno 2001; Van den Top 2003). CADC or CALC was allowed to be merged with CBFM providing an additional dimension and benefits of CBFM to CADC/CALT. It is important to note, though, that none of these instruments were able to achieve the complete devolution of power over forest resources, which are still under the ultimate supervision and authority of the state. The rights of indigenous people were further strengthened by Peoples' Right Act in 1997, which allows the granting of a Certificate of Ancestral Domain Title (CADT) and a Certificate of Ancestral Land Title (CALT) to recognized indigenous communities that were able to prove their land occupation since immemorial times. These certificates are issued by an independent state agency, the National Commission on Indigenous Peoples, conferring complete usufruct rights to indigenous communities, and thus represent complete forest devolution (Ballesteros 2001). Despite a long experience with these programs, so far no government report have been released on their outcome, and to date peer-reviewed literature has focused on governmental weakness (Ballesteros 2001; Dahal and Capistrano 2006; Pulhin and Dressler 2009; Pulhin 2003). As far as the author knows, no research has so far been carried out on the implementation of CBFM to indigenous communities and CADT or CALT in the context of forest resources. Hence, the present chapter attempts to cover this gap in the literature, by examining changes that have taken place in such communities after the implementation of the CBFM, and what problem are constraining their future development.

2 Methodology

2.1 Case Study in the Philippines

Two case studies, representing the communities which have participated in the CBFM program for approximately 15 years, were selected in the northern Philippines (Fig. 1).

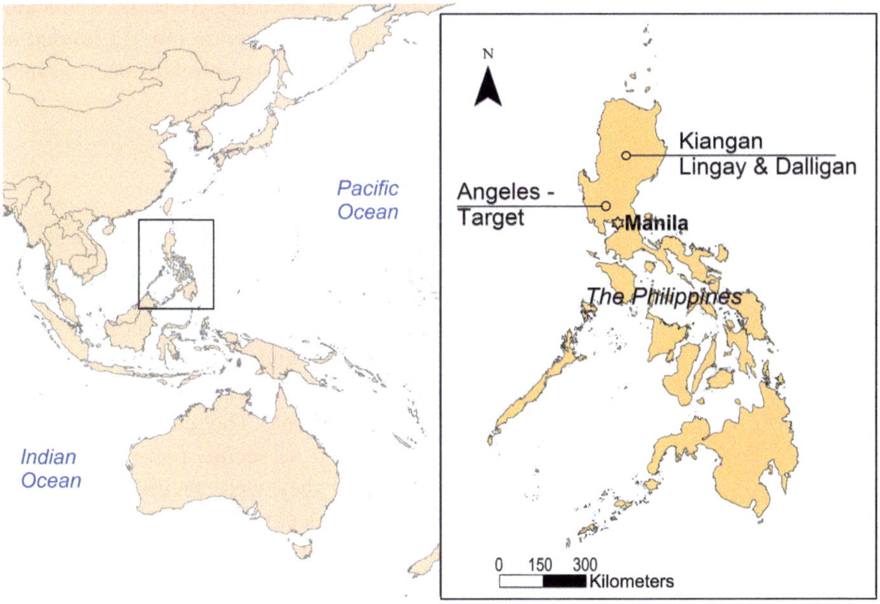

Fig. 1 Location of case studies

2.1.1 Kiangan—Cultural Landscape of Forest and Rice Terraces

Two remote districts in the Philippines, Lingay and Dalliga, in Kiangan municipality, Ifugao province, are inhabited by the Tuwali tribe of the Ifugao people. The forest system and communal or private management of the Ifugao province is regarded as viable, with many authors taking it as a model of sustainable community forestry based on customary law (Pulhin and Pulhin 2003; Perera 2009; Hlaing et al. 2013). These districts have a rather abrupt terrain, with the hills being an integral part of the landscape in an ecological and semantical way, which provide water for rice terrace irrigation, firewood, and timber for carving and construction. Two CBFM agreements were initiated by DENR, with a World Bank fund in 1999. Before the CBFM, a CADC was awarded to the area in 1996 to recognize the presence and practices of the indigenous community.

2.1.2 Target Village—Forest Dependent Semi-nomadic Tribe

The village of Target is located in the Angeles municipality, Pampanga province, with the majority of its inhabitants belonging to the Aeta people. The Aeta people have been recognized as being strongly dependent on local forest resources, for firewood, hunting and agroforestry (Reed 1904; Seitz 1998), and are considered to be spiritually connected to their forest (Shimizu 1989). The group was awarded two funds by the Japanese Bank for International Cooperation (JBIC) to implement a CBFM Program in 1998 for watershed reforestation and livelihood improvement, after the area was devastated in the early 1990s by a volcanic eruption. In 2004 the Japan International Cooperation Agency (JICA) funded an enhancement project for the CBFM. In 1997, the area had already been granted CADC, which in 2007 was transformed into CADT.

2.2 *Implementation of the Workshop*

The workshops conducted were constituted of both Participatory Mapping and Problem Ranking exercises. They were conducted in three indigenous communities, two in Kiangan (Ifugao Province)—Lingay and Dalligan, belonging to the same Tuwali tribe of the Ifugao people; and in the Target village of the Aeta people in Angeles municipality (Pampanga Province). The communities were informed about the workshop 2 weeks in advance to secure participation. The workshops were conducted in the year 2015, and they were facilitated by interpreters affiliated to a local NGO (Save Ifugao Terraces), Ifugao State University, and the University of the Philippines. A brief summary of each case study and the number of participants is presented in Table 1.

Table 1 Summary of case studies and workshop implementation

Municipality and Province	Village	Indigenous people	CBFM since	Workshop date	Number of participants[*]/ number of households in the community	Participants' background[**]
Kiangan, Ifugao	Lingay	Tuwali tribe of Ifugao people	1999 (CADC in 1996)	February 19, 2015	6/80	Formal education: none to high school Age: 30s to 80s
	Dalligan		1999 (CADC in 1996)	February 20, 2015	13/81	Formal education: none to college Age: 20s to 70s
Angeles, Pampanga	Target	Aeta people	1998 (CADC in 1997, CADT in 2007)	February 26, 2015	4/151	Formal education: none to college Age: 30s to 50s

[*]Heads of households
[**]All participants were original residents of each specific area

The consents of the indigenous people to make use of the data collected through the research were secured in all locations by complying with the requirements of the National Commission on Indigenous Peoples, in accordance to the Indigenous Peoples Rights Act of 1997. Moreover, local authorities were informed and consulted about the research to ensure transparency.

2.3 Conducting Participatory Mapping

Participatory Mapping was carried out to identify changes in a given community by drawing and describing maps "before" and "after", meaning the situation before CBFM implementation and at the time of the workshop. Participatory Mapping was conducted according to the process described in Mukherjee (2002), and Rietbergen-McCracken and Narayan-Parker (1998). The participants, local people with knowledge on the place, draw on a piece of paper a map of their communities and resources. In the process of mapping, key participants were responsible for actually drawing the map, while the other participants present provided active support. The Participatory Mapping started by deciding together with local people which type of map should be sketched, followed by the actual workshop with people who were familiar with the area and were willing to share their knowledge. Participants were provided with support whenever needed, but in general were left to draw the map by themselves. The mapping and discussion took approximately an hour and half in each study site. The output of the community work was questioned by the researcher, with all issues being noted and discussed.

Table 2 A sample of problem ranking matrix table

	Problem 1	Problem 2	Problem 3	Problem 4	Problem 5	Total
Problem 1						**0**
Problem 2	√					**1**
Problem 3	√	√		√		**3**
Problem 4	√	√				**2**
Problem 5	√	√	√	√		**4**

2.4 Conducting Problem Ranking

Problem Ranking exercises were carried out to identify issues that were part of the community's "bottle neck" in developing their desired livelihood strategy. Problem Ranking was conducted according to the guidelines by Rietbergen-McCracken and Narayan-Parker (1998). The participants were encouraged to list up to six main problems in their community (general or specified), and afterwards they were asked to rank them in order of importance, by showing two problems in two separate cards at once. The results were recorded in the form of a matrix for pairwise systematic ranking, using two cards at a time, asking participants to choose the one which represented the bigger problem. The number of times that each problem was selected when presented in such pairs gives the ranking of the most urgent problems in the community. A sample matrix is shown in Table 2, in which each problem in the first column is compared to the problem in the first row, and "ticked" when it is the more important one. The sum of the number of "ticks" in the right-hand side column forms the basis of the final ranking order, where the most important problem is considered to be the one which has the highest number of "ticks" (with the least important problem being the one with the lowest number).

3 Results

3.1 Kiangan

Participatory Mapping and subsequent discussions were able to provide a picture of the changes that have taken place in both Lingay and Dalligan, which have essentially undergone similar changes over the past few decades. An example of the maps that were drawn in these locations is presented in Fig. 2. Initially both districts were unified as one, called Lingay. However, due to population growth a remote part of the district was separated from the main Lingay, and named Dalligan.

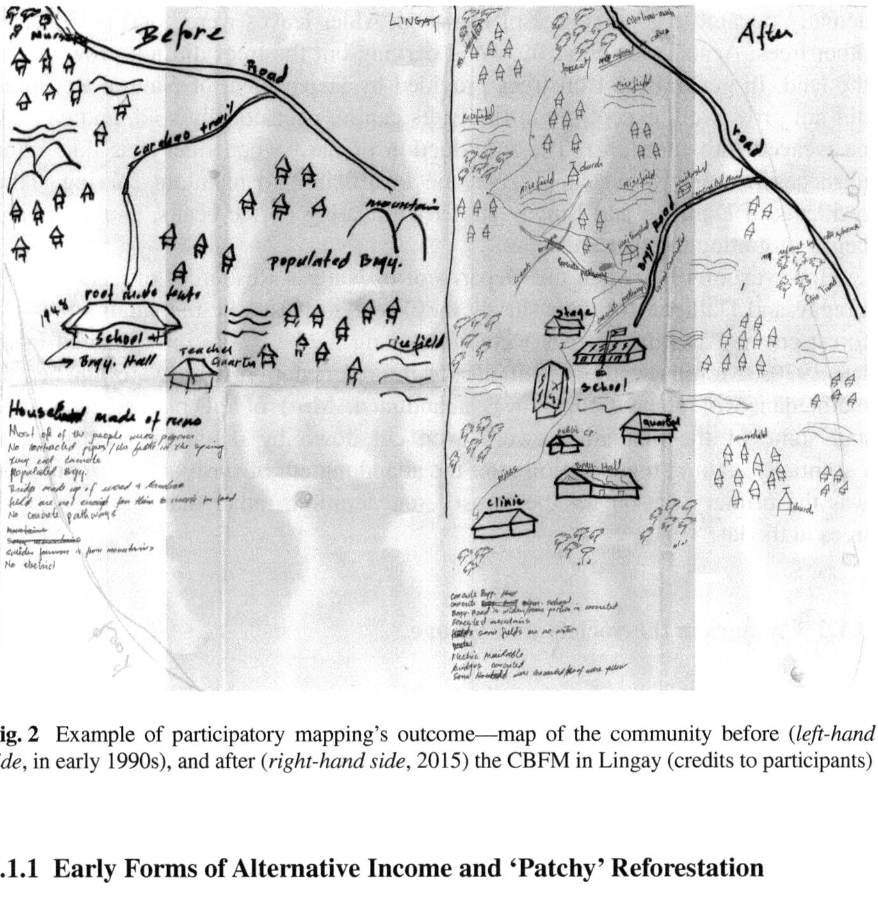

Fig. 2 Example of participatory mapping's outcome—map of the community before (*left-hand side*, in early 1990s), and after (*right-hand side*, 2015) the CBFM in Lingay (credits to participants)

3.1.1 Early Forms of Alternative Income and 'Patchy' Reforestation

Due to intensive slash-and-burn practices mountain slopes were bare by the late 1960s. In 1968, wood carving started in the area as a source of income. Wood carvers and blacksmithing knowledge came from another district. Thanks to these new skills the residents in Lingay and Dalligan could make new products and exchange them for rice. As a consequence of introducing these alternative livelihoods, fewer slash-and-burn practices were needed to produce alternative staple foods such as sweet potatoes. Slash-and-burn practices continued to decrease—being abandoned in the 1970s- and gradually the forest regenerated.

In 1979 the Department of Natural Resources started introducing a number of trees into the area, such as gemelina, pine trees, mahogany, alder, and acacia, though these turned out to be generally unsuitable to the local environment. According to the participants, gemelina should not be planted near a source of water, as its roots absorb large quantities of water and other trees could not grow around it. Gemelina was found to be suitable for furniture but wooden nails (as opposed to those made of steel) had to be used to avoid rusting, as gemelina wood contains much water. The best types of trees for the area were found to be alder and mahogany. Mahogany, which was intended to be a source of hardwood, actually became a good source of firewood. Alder leaves were good fertilizer for other trees. Acacia was good for wood carving but the trees did not grow well in the land. In general the fruit trees provided by Department of Natural Resources did not grow well, except rattan (its fruits can be occasionally sold, and can also be weaved into furniture). The introduction of hard wood trees was thus rather unsuccessful, and wood supply became insufficient to maintain carving. Thus, residents of Dalligan and Lingay started cultivating coffee beans, and afterwards began migrating to other areas.

However, in 1986 when the Deparment of Natural Resources officers came to Lingay and Dalligan for their survey they were surprised by the small amount of forest cover. Essentially, trees were still young as they were only planted in the late 1970s. Moreover, after the funds for the reforestation work were finished the maintenance of planted forest was abandoned. Most of the planted trees died, and some of the plantations were even cut down by dissatisfied participants. Essentially, natural regeneration after the abandonment of slash-and-burn practices was the primary reason for the forest's regeneration, rather than the planting of trees in the late 1970s.

3.1.2 Changes in the Society and Village

The diet of people has been gradually changing since the 1960s from one of sweet potatoes and rice from their local farms to predominantly non-native rice purchased in the town, with money earned from selling wood carvings, furniture, coffee, rattan, or temporary income from working on reforestation projects. Another change that took place in Lingay and Dalligan was Christianization during the 1980s (before that time people believed in the Ifugao gods), and since then many churches have been built. The gradual change in religious practices caused a reduction in rituals associated with rice planting and the forest. Other developments that occurred in recent times in Lingay were electrification, the building of a clinic, a concrete bridge replacing the wooden one, a concrete local district hall and teacher's quarter, and the school being rebuilt using concrete. Nowadays, wooden carving has disappeared due to the high cost of the hardwood, its lower availability, and difficulty in transportation. The housing patterns have also changed, from traditional houses covered with thatched roofs, to metal sheet roofs, with some of the houses already being made of concrete. At present there are already concrete and metal pipes connecting houses with local springs, and some people are able to take a bath in their homes instead of in the spring.

3.1.3 No Changes Through CBFM

In 1999, Dalligan and Lingay were once again (i.e. after the first projects 30 years earlier) included into a reforestation program, this time under CBFM. Similarly to the former project, the CBFM provided only temporary employment through paid forest planting. Participants received seedlings that were used for reforestation, planting communal orchard, and for private woodlots. Plantation of fruit trees as well as planted forest was abandoned after the reforestation funds were finished. Therefore, only the planted forest tree species were able to improve private forest lots, but not the common plantation. Moreover, low rates of planted seedlings' survival were reported in common areas due to lack of maintenance. This lack of incentives to provide care to young forest plantation was similar to the failure that was recorded during the reforestation program in the late 1970s. However, compared to the past, forest cover was already higher, and an increase in biodiversity could be noticed, e.g. respondents reported how monkeys had returned to the forest. The implementation of CBFM was not a very significant event in these communities, and thus participants of the workshop had difficulty in recalling the implementation of the CBFM. However, CBFM promoted livestock, and prior to it local inhabitants had kept native pigs for private rituals and consumption, though at present they are raising hybrid pigs for selling. Animal feed has also changed, from sweet potatoes and other root crops that the native pigs had been fed with to commercial food for hybrid pigs.

3.1.4 Problems in the Communities

Major problems in the communities were related to having no road connection and discontinuities in the irrigation or drinking water system. Compared to the past, nowadays the road was wider, with some part covered with concrete. However, it is generally still not good enough to be safely transited during rainy seasons, when its poor condition sometimes prevents locals from accessing the market, where they could potentially sell their products and purchase food.

The problem of some rice paddies being without water was caused by a lack of maintenance of the rice terrace system. Once some of the rice fields were abandoned or stop being maintained, it triggered rice terraces below them to be cut off from the water. Such abandoned rice fields created a haven for pests such as rats, which can reduce rice yields in adjacent fields. The problem of the maintenance of potable water pipes, listed during the workshop, was also caused by a lack of collective action between all households to maintain the system. A shortage of some facilities, such as health centers and schools, were also pointed out in Dalligan as being important. All these problems are listed in a hierarchical order in Table 3.

Table 3 Problems ranked by villagers during workshops

Kiangan		Angeles
Lingay	Dalligan	Target
1. Drinking water supply 2. Farm to market road 3. Irrigation 4. Source of income 5. Lack of food supply (esp. rice) 6. Reforestation (boundary dispute)	1. Road connection 2. Irrigation for livelihood 3. Health center 4. Drinking water system 5. Facilities in the school	1. Malnourishment 2. Lack of toilet 3. Access to education 4. Solid waste management 5. Noise and drinking

3.2 *Target*

3.2.1 From Semi-nomadic Group to a Sedentary Community

In the 1960s and 1970s only six houses of the Aeta people were in what is currently the Target village, which was mostly composed of buildings belonging to the American army, who had a base in the area. In fact, the village of Target used to be an area where the American military conducted practice shooting, which is at the origins of the village's name—Target. At that time the Aeta were semi-nomadic, and their population was spread over the adjacent mountain fields. Houses were constructed using traditional materials, with walls made of bamboo, and roofs thatched with cogon grass. In 1990, the year before Mt. Pinatubo eruption, a map drawn during a community mapping workshop depicted already 50 houses at the current location of Target. Residents were mostly Aetas, with some people coming from outside the community. The community had a communal farm (present till now). The road which ran through the settlement had a different layout from the current one. Some areas adjacent to the settlement were not owned by the Aeta people but a private company, which generated a problem of land access for the Aeta. At that time the surrounding forest was in a very good condition, which was used for hunting and food gathering by local people. Work in the forest and farms (mostly shifting cultivation type farms) were facilitated by cooperation among community members, who planted mostly corn and yum. A culture of common use of resources was still deeply rooted in people's mind, and thus, anything grown in the mountains was considered as common goods. At that time the main source of water was a spring in the mountains, which was fetched and transported via water tanks to the village. Meanwhile the Aeta people also started converting from their tribal beliefs to Christianity as a result of the work of a local Baptist church.

3.2.2 Recovery of the Community and Forest

In Target the critical event that has marked the Aeta people was the eruption of Mt. Pinatubo in 1991. The volcanic eruption caused a temporary evacuation of the inhabitants from the area, after which people had to re-start their lives, rebuild their houses and recover farms. In the village, 50 houses were built by a private foundation. Target also gained an elementary school. Water, instead of being transported from the spring in the mountains, started to come from artesian wells perforated in the community and then transported to the houses in water tanks (though it appears that such water occasionally caused sickness amongst residents). A new road through the place was laid, with other houses being built in the traditional way. However, the new road was not paved, making transportation difficult. In the new Target, the previous church was replaced by a Born Again church, funded by Korean donor., This church became very influential in the group, bringing new facilities, and actively promoting Christianity in the community, and this brought further damages to the tribal beliefs. Communal farms and the rough roads that connected them continued to be used in the same way, though in 2004, a spa resort was built in the mountains, together with a restaurant (which opened in Target itself). The spa and the restaurant became places where local residents could obtain permanent employment, and this reduced participation in communal and individual farming.

3.2.3 Ownership Issues and CBFM Implementation

The CBFM Project provided support to plant trees after the volcanic eruption, and thus nowadays there are more trees than immediately after the disaster. However, trees are still rather young compared to the natural thick forest before the disaster. Land ownership also changed from communal ownership, in which people could change their farm location in the process of shifting cultivation, to permanent occupation of specific areas. The decision to change the land ownership was taken by community elders to protect their land from outsiders, who began acquiring land titles. Despite the changed mode of land ownership, Aeta people can still lease land among themselves if needed. Also, the cultural custom of sharing crops and fruits with anybody who wants to eat is still present, and such cultural values have been passed to the younger generation, according to the participants of the workshop. In 2007 the community was awarded with CADT. CADT area did not cover the village of Target or its adjacent area, but the mountainous area where most of the Aeta's farms are located.

The outcome of the mapping exercise in Target is presented in Fig. 3.

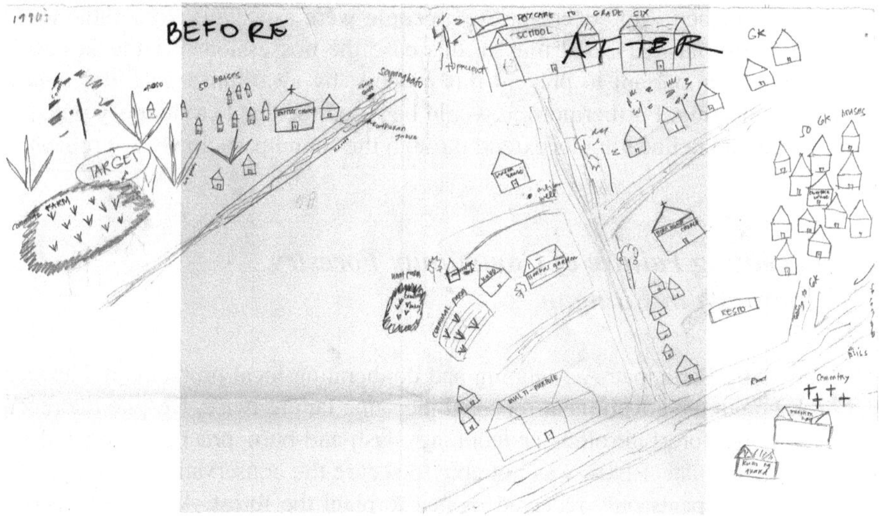

Fig. 3 Participatory mapping's outcome in Target—map of the community before the volcanic eruption (*left-hand side*) and after the CBFM Project implementation, as of 2015 (*right-hand side*) (credits to participants)

3.2.4 Problems of Settlement

The problems that Target is struggling with are mainly related to the malnourishment of children, sanitary and waste facilities in the houses and village, access to post-elementary education, and peace and order in the village (Table 3). Since Aeta begun working together with non-Aeta residents in the resort, many of young men started drinking in the evening, causing sporadic troubles.

4 Discussion

4.1 CBFM Program Potential Role

Through discussions with participants over the results of Participatory Mapping and problems that the communities had, the CBFM Project was found to have made an important contribution to the community, by helping to secure the production or purchase of food and to assist reforestation. However, not all issues were adequately addressed. Participants, through the mapping process, were not able to clearly define the extent of the CBFM area. It was originated in the fact that each respective area had its own local ownership system that was not addressed by the CBFM policy. Thus, the CBFM program was associated with reforestation activities. It implies a very weak understanding of the nature of the CBFM program, which is a tool for granting access to land, including forests, within specific boundaries. Only small lots in Kiangan were recognized as communal plantations, but were abandoned after CBFM funds for reforestation work were finished. Moreover, the Aeta people were especially vulnerable to the takeover of by outsiders, and thus in this case the possession of either a CBFM or CADT was important to prevent this process, by awarding legal titles regarding land ownership. Furthermore, it would be recommendable to identify potential CBFM area and delineating its extend through the sound participatory approach.

4.2 Repeating Failure of Community Forestry Approach in Kiangan

By conducting Participatory Mapping and deliberating local problems it was clear that in Kiangan the government repeated the same failure twice. Despite the regeneration of the forest through abandoning slash-and-burn practices, the CBFM, which started in late 1990s, was not able to secure the conservation of the planted forest, as participants only received money to plant the forest. Afterwards, people became absorbed by meeting basic needs, such as rice production and refused to work to maintain the planted forest, though the extensive slash-and-burn practices did not revive (as people were already able to generate minimum income from various other resources). Furthermore, many essential problems in these villages, such as the maintenance of irrigation or water systems to provide potable water to the households (which required communal work), were not addressed by the program. Additionally, the community-based management of rice terraces (that became abandoned) was not taken into consideration, and this only made local problems worse, especially when people who had not intended to abandon their rice paddies were forced to do so (for example, due to separation from the irrigation system, as some farmers abandoned their fields, creating gaps in the network). Moreover, the problem of lack of access to a market was a major constraint for selling local products, and thus the promotion of market-based livelihood was bound to fail.

4.3 Mediocre Outcomes of CBFM Project

Through the workshop in Target, it could be understood that farm-based livelihoods and reforestation were enhanced to a certain extent by the CBFM Project. The group was able to access the market to sell their product, though the poor road still hindered their access to the market. The distance from the village to their own farms was another difficulty that local farmers had to overcome. Most of the land surrounding the settlement was not Aeta property, whereas the CBFM project area was located in the distant mountains. In that sense, the CBFM or CADT was necessary to secure a minimal area so that the Aeta could preserve their livelihoods, but was not able to recover the area already occupied by outsiders (either private companies or individuals).

Local problems that were found in Target were related to those of a sedentary lifestyle, which was still a new way of life compared to their traditional semi-nomadic characteristics. Also, interactions with non-indigenous people who began to live in the same village were a source of local problems. The commodification of their sweet potatoes, which in the past were a valuable nutritious component of their diets, has currently become a product to generate income. The income from such products is nowadays used for the purchase of rice and processed food, and the changes in diet could be a cause of some of the community's problems—for example leading to the malnourishment of children, but this would require further research to be clearly established. Thus, the workshop showcased that the local problem identification could be the first step to envision possible future solutions prior the projects.

4.4 Development Paths of Indigenous Communities

People in the municipality of Kiangan were found to be preserving their traditional lifestyle, which remained localized and heavily dependent on natural resources, and CBFM did not introduce any major changes in livelihood strategy.

The community was able to minimize the slash-and-burn practices for food production in the local forest, though further threats to local sustainability have arisen due to problems deriving from the discontinued use of rice terraces and unmaintained irrigation and drinking water systems. Such problems probably require intervention to ensure long-term sustainability.

In the village of Target CBFM was not able to change the localized strategy of the community, which meant it still maintained its tradition of common property and livelihood still depended on local resources (though such traditions were weaker than what they once were). CBFM was not able to increase income due to low pricing of crops, though it promoted the commodification of farm products. However, the establishment of a spa resort offering regular employment is likely to increase the pressures of globalization and act as a catalyst for rapid changes. It was evident that a number of social problems have appeared within the community because of new patterns of occupations, which had not existed before the volcanic disaster. To ensure the long-term sustainability of this community, it would also be necessary to solve problems related to waste management and potable water, and to guide farmers into diversifying their food production to secure a balanced diet.

The localization strategy is very resilient (2012), and in case of indigenous people developed on the social structure, culture, and available natural resources as the source of subsistence (Abel et al. 2006). The case studies of CBFM represented that decentralization of the forest was not effective instruments for enabling alternative development paths. Community forestry was able to boost resilience during the initial stage of the program through income generation, but after the funds allocated for implementation ran out the group returned to their previous forms of livelihood, thus having no impact on their livelihood strategy. Thus, the recommended for collaboration of stakeholders and collective management of resources glocalization strategy is yet to be developed (Inoue et al. 2015), but more comprehensive and applicable form of community forestry policy needs to be crafted.

4.5 Participatory Appraisal Experience

The two method of the participatory appraisal applied to the research, the Participatory Mapping and the Problem Ranking, demonstrated that there is certain level of usefulness of these methods in the sustainability research as well as in the practice of implementation of sustainable development projects. Chambers (1994) elaborated in his review study that participatory appraisal had the capacity to perform with high validity, and produce reliable data compared to traditional methods. In such case, high reliability of the information shared by the participants is crucial. In this context, Campbell (2001) highlighted that these methods may raise issues of validity and reliability causing more problematic use such participatory appraisal in sound research, but nonetheless able to produce roughly accurate information, useful for explorative stage of the research. Thus, this method can compromise long, stationary in Field-based, with short, visit to the field-based, research, producing data coming from local people in relatively short time, and that can be beneficial for describing local context or for parishioners of sustainable development aiming specific community as a target of development project (Chambers 1997).

4.5.1 Participatory Mapping's Advantages and Limitations

In the light of listed issues, the implementation of the workshop generated a number of practical advices. Facilitation but minimum intervention of the research into participatory mapping was the key point to encourage the participants into drawing on the paper their own knowledge on the place and share it during discussion. In the first part of the workshops, Participatory Mapping exercises were conducted. The author of the chapter would like to note how hesitant at first some of the participants were to sketch the map, as they had little experience in drawing. Therefore, the facilitation of the map drawing process was very important, by observing the members and, if necessary, encouraging them to continue with the exercise. Essentially, it was important to appoint one main person to draw the map, with others being mostly engaged in the discussion and giving further directions for the map. Discussions during mapping and after finalizing the map revealed a number of events and facts that were happening in each community, which allowed further discussions on the community development and the CBFM project implemented in these communities. Limitations of the method is use of the paper and pen, which may not always be the most accurate way of representing precisely spatial conditions, such as area, distance, etc. Thus, for more accurate data on local land use and past changes, further study with use of GIS software, satellite images, and available maps is essential.[2]

4.5.2 Problem Ranking's Advantages and Limitations

Raising engagement and interests in the workshop among participants accelerates further participation during the workshop, maximizing the knowledge on the local problems and ideas on causes or solutions. Problem Ranking was found to be an engaging exercise whereby people could discuss existing problems in their communities. As a tool, it would appear to sit at the edge between being a workshop and a game, in which ranking the problems they listed involved discussion and excitement among participants. The final outcome of the Problem Ranking drew the attention of participants, as they were interested in the final score of their voting and ranking process. The author found this method to be very entertaining for participants, far more than Participatory Mapping. As a suggestion to other

[2]Further study of presented areas, with use of GIS methods is yet to be published.

researchers, it could thus be recommendable to start the workshop with Problem Ranking, and then move on to Participatory Mapping. What is also noteworthy was how the tool was modified during the workshop, regarding the situation where two of the problems were ranked with the same level importance. This happened in Lingay, and it was necessary to make an additional voting to decide the most import problem within the pair, to fix the final hierarchical order.

5 Conclusion

In this chapter the methodology of Participatory Mapping and Problem Ranking were explained and applied to two case studies of communities in the Philippines which participated in CBFM Projects. The tools were able to engage participants in the research during the workshop, as well as provide entertainment to the while interacting with the researcher. However, it is important to note that careful observations and support are needed throughout any given workshop to keep the participants active.

These tools were able to depict the local context and problems being faced by indigenous people, and to rapidly assess the development programs outcomes, identifying strategies that the community desired to take in the future. Both communities were found to be following localized development paths, and local problems were either hindering changes or weakening the strategy. In Kiangan the geographical remoteness of the villages was still delaying the changes. However, in Target, where the remoteness of the area was not the major problem, future globalization impacts are more likely to arise due to changes in lifestyle (to a more sedentary nature) and the introduction of permanent employment.

The CBFM program in the Philippines was found to have an incomplete approach to reforestation, repeating failures from the past. Also, attempts to improve local livelihoods failed, as they were not properly addressing some of the major problems in the villages, which were constraining the sustainability of CBFM in the long term. As the next step to achieve the goals of the CBFM policy a revision of the policy regarding its implementation and long-term maintenance is needed, especially in addressing complex local situation and problems, and the issue of ill-defined boundaries within project forests.

Acknowledgements I would like to express deepest thanks and appreciation to communities of Lingay, Dalligan, and Target, for participation in the workshop and sharing the knowledge. I also thank the Graduate Program in Sustainability Science-Global Leadership at The University of Tokyo for financial support in funding field study, and to editors of the book for insights and comments on the chapter.

References

Abel, N., Cumming, D., & Anderies, J. (2006). Collapse and reorganization in social-ecological systems: Questions. *Some Ideas, and Policy Implications, 11*(1), 17.

Ballesteros, A. (2001). Ancestral domain recognition and community based forest management: Two sides of different coins. *Philippine Natural Resources Law Journal, 11*, 35–61.

Bavinck, B., & Jyotishi, A. (Eds.). (2015). *Conflict, negotiations and natural resource management: A legal pluralism perspective from India*. Oxon and New York: Routledge.

Bryan, J. (2011). Walking the line: Participatory mapping, indigenous rights, and neoliberalism. *Geoforum, 42*, 40–50. doi:10.1016/j.geoforum.2010.09.001.

Bryan, J. (2015). Participatory mapping. In T. Perreault, G. Bridge, & J. McCarthy (Eds.), *The Routledge handbook of political ecology* (pp. 249–262). Oxon and New York: Routledge.

Campbell, J. R. (2001). Participatory rural appraisal as qualitative research: From participatory claims. *Human Organization, 60*, 380–389.

Chambers, R. (1994). Participatory rural appraisal (PRA): Analysis of experience. *World Development, 22*, 1253–1268. doi:10.1016/0305-750X(94)90003-5.

Chambers, R. (1997). Shortcut and participatory method for gaining social information for projects. In Sustainable Development: S. Sepulveda, R. Edwards (Eds.), *Social organization, insitutional arrangements and rural development: Selected readings* (vol. 6). Inter-American Institute for Cooperation on Agriculture.

Dahal, A., & Capistrano, D. (2006). Forest governance and institutional structure: An ignored dimension of community based forest management in the Philippines forest governance and institutional structure: An ignored dimension of community based forest management in the Philippines. *International Forestry Review, 8*, 377–394.

Food and Agriculture Organization. (1992). *CF note 7: Community forestry: Ten Years in review*. Rome: Food and Agriculture Organization of the United Nations.

Fraser, E., Dougill, A., Mabee, W., Reed, M., & McAlpine, P. (2006). Bottom up and top down: analysis of participatory processes for sustainability indicator identification as a pathway to community empowerment and sustainable environmental management. *Journal of Environmental Management, 78*, 114–27. doi:10.1016/j.jenvman.2005.04.009.

Hlaing, E., Inoue, M., & Pulhin, J. (2013). A property-rights approach to understanding regulations and practices in community-based forest management: Comparison of three systems in the Philippines. *Small-scale Forestry, 12*, 579–596.

Howitt, R., & Suchet-Pearson, S. (2006). Rethinking the building blocks: Ontological pluralism and the idea of "management". *Geografiska Annaler: Series B, Human Geography, 88*, 323–335.

Inoue, M. (2003). Participatory forest management policy in South and Southeast Asia. In M. Inoue & H. Isozaki (Eds.), *People and forest—policy and local reality in Southeast Asia, the Russian Far East, and Japan* (pp. 49–72). The Netherlands: Kluwer Academic Publishers.

Inoue, M., Shivakoti, G., & Ojha, H. (2015). Multi-level Forest Governance in Asia: An Introduction. In M. Inoue & G. Shivakoti (Eds.), *Multi-level forest governance in Asia—concepts, challenges and the way forward*. New Dehli: SAGE Publications India.

Levine, A. S., & Feinholz, C. L. (2015). Participatory GIS to inform coral reef ecosystem management: Mapping human coastal and ocean uses in Hawaii. *Applied Geography, 59*, 60–69. doi:10.1016/j.apgeog.2014.12.004.

Magno, F. (2001). Forest devolution and social capital: State-civil society relations in the Philippines. *Environmental History, 6*, 264–286.

Mukherjee, Neela. (2002). *Participatory learning and action: With 100 field methods*. New Delhi: Concept Publishing Company.

Narayanasamy, N. (2009). *Participatory rural appraisal: Principles, methods and application*. New Delhi: SAGE Publications India Pvt Ltd.

Napoleon, V. (2013). Thinking about indigenous legal orders. In R. Provost, & C. Sheppard (Eds.), *Dialogues on human rights and legal pluralism*. Dordrecht: Springer, pp. 229–426.

Perera, Jaysntha (Ed.). (2009). *Land and cultural survival: The communal land rights of indigenous peoples in Asia*. Metro Manila: Asian Development Bank.

Platteau, J., & Abraham, A. (2002). Participatory development in the presence of endogenous community imperfections. *Journal of Development Studies, 39*, 104–136. doi:10.1080/00220380412331322771.

Pulhin, J. (2003). Trends in forest policy of the Philippines. In M. Inoue, & H. Isozaki (Eds.), *Policy trend report 2002* (pp. 29–41). Tokyo: The Institute for Global Strategies (IGES) Forest Conservation Project. Soubun Printing Co., Ltd.

Pulhin, J., & Dressler, W. (2009). People, power and timber: the politics of community-based forest management. *Journal of Environmental Management, 91*, 206–14.

Pulhin, J., Peras, R., & Tapia, M. (2015). Philippines: Multi-tired forest governance system on uneven playing ground. In M. Inoue & G. Shivakoti (Eds.), *Multi-level forest governance in Asia—concepts, challenges and the way forward* (pp. 227–256). New Dehli: SAGE Publications India.

Pulhin, J., & Pulhin, P. (2003). Community-based forest management in the Philippines: Retrospect and prospect. In M. Inoue & H. Isozaki (Eds.), *People and forest—policy and local reality in Southeast Asia, the Russian far east, and Japan* (pp. 139–156). Dordrecht: Kluwer Academic Publishers.

Reed, William. (1904). *Negritos of Zambales*. Manila: Bureau of Public Printing.

Reyes-García, V., Orta-Martínez, M., Gueze, M., Luz, A. C., Paneque-Gálvez, J., Macía, M. J., & Pino, J. (2012). Does participatory mapping increase conflicts? A randomized evaluation in the Bolivian Amazon. *Applied Geography, 34*, 650–658. doi:10.1016/j.apgeog.2012.04.007.

Rietbergen-McCracken, Jennifer, & Narayan-Parker, Deepa. (1998). *Participation and social assessment: Tools and techniques* (Vol. 1). Washington: The World Bank.

Robinson, J., Burch, S., Talwar, S., O'Shea, M., & Walsh, M. (2011). Envisioning sustainability: Recent progress in the use of participatory backcasting approaches for sustainability research. *Technological Forecasting and Social Change, 78*, 756–768. doi:10.1016/j.techfore.2010.12.006.

Seitz, S. (1998). Coping strategies in an ethnic minority group: The Aeta of Mount Pinatubo. *Disasters, 22*, 76–90.

Shimizu, H. (1989). *Pinatubo Aytas: Continuity and Changes*. Quezon City: Ateneo de Manila University Press.

Shivakoti, G., Inoue, M., Pulhin, J., Sharma, S., Webb, E., & Nath, T. (2015). Towards an effective policy for forest management in Asia. In M. Inoue & G. Shivakoti (Eds.), *Multi-level forest governance in Asia—concepts, challenges and the way forward*. New Dehli: SAGE Publications India.

Tobin, B. (2014). Indigenous peoples, customary law and human rights—why living law matters. Oxon & New York: Routledge.

UNSECO. (2008). Operational guidelines for the implementation of the world heritage convention. UNESCO World Heritage Centre. Retrieved January 2, 2016 from http://whc.unesco.org/en/guidelines.

Van den Top, G. (2003). The social dynamics of deforestation in the Philippines: Actions, options, and motivations. Copenhagen: Nordic Inst of Asian Studies.

Veland, S., Howitt, R., Dominey-Howes, D., Thomalla, F., & Houston, D. (2013). Procedural vulnerability: Understanding environmental change in a remote indigenous community. *Global Environmental Change, 23*, 314–326.

Vergragt, P. J., & Quist, J. (2011). Backcasting for sustainability: Introduction to the special issue. *Technological Forecasting and Social Change, 78*, 747–755. doi:10.1016/j.techfore.2011.03.010.

Viano, C. E. (2015). The curse of Riches: Sharing Nature's wealth equitability? In J. Wouters, A. Ninio, T. Doherty, & H. Cisse (Eds.), *The World bank legal review volume 6: Improving delivery in development: The role of voice, social contract, and accountability*. Washington: The World Bank.

Wilson, G. (2012). *Community resilience and environmental transitions*. New York: Routledge.

Ziai, Aram. (2015). *Development discourse and global history: From Colonialism to the sustainable development goals*. Oxon and New York: Routledge.

Rapid Sustainability Appraisal of Collapsed Jatropha Projects in Ghana Using Local Community Perceptions: Methodological Implications for Sustainability Science

Abubakari Ahmed and Alexandros Gasparatos

Abstract In the mid-2000s, Ghana experienced a biofuel boom driven by jatropha expansion, but within a few years almost all jatropha projects within the country collapsed. Limited community participation in biofuel project planning has played a key role in the failure of the projects, which have had a range of sustainability impacts on the local environment and society. Understanding the patterns of sustainability impacts through community perceptions can provide information that could enhance the sustainability of future biofuel projects in the country. By using a rapid sustainability appraisal that captures community perceptions, the present chapter compares the sustainability impacts experienced by communities around three failed jatropha projects in Ghana with those captured in their respective Environmental Impact Assessment (EIA) reports. The authors found a mismatch between the impacts experienced by communities and those assessed in EIAs that arises, to an extent, from the limited participation of communities in the project planning and EIA processes. The findings suggest the need for adopting a bottom-up approach for the identification and selection of sustainability impact criteria. Sustainability science scholars can use rapid sustainability appraisals to gain an initial understanding of a given study area, in order to inform the framing of the research questions and help in the selection of an appropriate methodology to collect and analyse actual data through subsequent fieldwork.

Keywords Biofuel · Community participation · Environmental impact assessment · Jatropha · Project collapse · Rapid appraisal · Ghana

---

A. Ahmed (✉)
Graduate Program in Sustainability Science-Global Leadership Initiative,
Graduate School of Frontier Sciences, The University of Tokyo, 5-1-5 Kashiwanoha,
Kashiwa, Chiba 277-8563, Japan
e-mail: abubakari.ahmed@s.k.u-tokyo.ac.jp

A. Gasparatos
Integrated Research Systems for Sustainability Science (IR3S), The University of Tokyo,
Tokyo, Japan

M. Esteban et al. (eds.), *Sustainability Science: Field Methods and Exercises*,
DOI 10.1007/978-3-319-32930-7_10

1 Introduction

Between 2000–2011 several countries in Sub-Sahara Africa (SSA) established biofuel[1] targets with different combinations of feedstock options (e.g.. jatropha,[2] oil palm, sugarcane) and production modes (smallholder/outgrower schemes, large plantations) (Gasparatos et al. 2015). Biofuel policies in SSA were largely put in place to promote energy security and rural development (Gasparatos et al. 2015; Mitchell 2011; Johnson and Silveira 2014) rather than help with climate change mitigation, as was the case in some developed countries (Gunatilake et al. 2014; Månsson et al. 2014; Gasparatos et al. 2013a).

While some countries in Africa such as Malawi have implemented biofuel programmes since the 1980s (Johnson and Silveira 2014), it was only after the mid-2000s that feedstock production for biofuels became an attractive avenue to attract foreign direct investments (FDIs) across SSA (de Schutter 2011a; de Schutter 2011b; Gasparatos et al. 2015; German et al. 2013; von Maltitz et al. 2014). This was largely a reflection of the expectations raised by the development of large international biofuel markets, particularly in the European Union. The perceived potential to export feedstock/biofuels to international markets provided a strong incentive to expand feedstock production, particularly jatropha, in many SSA countries (Schoneveld 2014). As a result, large-scale land acquisitions for plantation-based feedstock production became popular in countries as diverse as Ghana, Madagascar, Mozambique, Tanzania and Zambia (Ahmed et al. 2015; von Maltitz et al. 2014; Setzkorn and von Maltitz 2013).

Even though expansion was rapid (Schoneveld 2014), research has given mixed results regarding which mode of production is more sustainable (van Eijck et al. 2014; Romijn et al. 2014; von Maltitz et al. 2014). In fact, most feedstock plantations, and especially those in West Africa, collapsed within the first 5 years of operation (Ahmed et al. 2015; Schoneveld and German 2014; Schoneveld 2014). In Ghana, for example, out of the 34 biofuel plantation projects that started after 2005, only one project was operational as of the time of writing this chapter (January 2016). The remaining have collapsed or suspended operations for a number of reasons including: (i) bad financial planning, (ii) lack of continuous investor funding, (iii) mismanagement, (iv) poor site/feedstock selection, (v) community agitation, (vi) negative reportage from civil society organizations and (vii) unconstructive involvement of traditional (e.g. chiefs) and formal institutions (e.g. government ministries, Lands Commission) (Ahmed et al. 2015).

---

[1]For the purpose of this chapter, we refer to biofuels as liquid fuels (e.g. bioethanol, biodiesel, straight vegetable oil) derived from biomass through biochemical processing. The main biofuel end-uses include transport, cooking and rural electrification (Gasparatos et al. 2015).

[2]Jatropha is an oilseed crop that can be used to produce vegetable oil. While jatropha oil is toxic to humans and animals (Devappa et al. 2010), it can be used as fuel, either as straight vegetable oil or after being transformed into biodiesel (Achten et al. 2007).

Inappropriate project planning and collapse has been shown to substantially affect local communities in Ghana, and a number of consequences have been reported as a result, including the loss of agricultural land, access to ecosystem services, local livelihoods and food security (Boamah 2014a, b; Boamah 2010, 2011; Kidido and Kuusaana 2014; Schoneveld and German 2014; Schoneveld et al. 2010, 2011). Such negative effects, combined with fears of losing land tenure or land grabbing, have fuelled skepticism about the actual potential of jatropha to alleviate rural poverty or enhance national/local energy security.

In Ghana, the limited participation of local communities in biofuel project design and implementation can be seen as both a consequence of inadequate corporate commitment to actually benefit local communities, as much as a failure of formal and traditional institutions (Ahmed et al. 2015). In other parts of SSA such as Tanzania, the lack of community participation is noted as a major reason behind the failure of many biofuel projects (Hashim 2014). Even in Mozambique, where multi-stakeholder structures were put in place to enhance the sustainability and viability of biofuel projects, this happened only after the boom and bust of the jatropha sector (Schut et al. 2014; Slingerland and Schut 2014). Aside from biofuels, other renewable energy projects have collapsed in different parts of the world due to setbacks originating from participation issues, or have conversely succeeded where there was wider community participation (Aitken 2010; Devine-Wright 2007; Hinshelwood 2001; Walker 2008; Walker et al. 2010).

Community participation in project design could provide an avenue to incorporate community values in biofuel projects, possibly enhancing their long-term viability and sustainability (Hashim 2014). Additionally, improving community participation in renewable energy projects' design could strengthen social support and acceptance, build commitment to enhance project sustainability and provide a sense of ownership (Aitken 2010; Cass et al. 2010; Devine-Wright 2007; Hammami et al. 2016; Rahut et al. 2015; Selman 2004; Walker 2008; Walker et al. 2010) (Sect. 2.1). Active community participation could further help to clearly assign responsibilities between investors and communities (Hashim 2014), and thus build a good interface to enhance community-investor relationships (Sect. 2.1).

However, research on community participation in biofuel projects is generally limited to issues of public, social or community acceptance (Chin et al. 2014; Clancy et al. 2011; Cowell et al. 2011; Dragojlovic and Einsiedel 2015; Giannoccaro and Berbel 2012; Longstaff et al. 2015; Rosso-Cerón and Kafarov 2015). Some scholars have also addressed issues of smallholders' decision making regarding whether to participate in new biofuel projects, especially smallholder-based projects (Jumbe et al. 2009; Kuntashula et al. 2014; Montefrio et al. 2015). Such studies, however, have not dealt with the need to integrate the potential sustainability impacts of biofuel expansion experienced by local communities, during project design.

The aim of this chapter is to undertake a rapid appraisal of the sustainability impacts of three collapsed jatropha projects based on the perceptions of local communities. The proposed methodology is based on the concepts of community participation and social acceptance (Sect. 2.1). The results of this study highlight how a rapid appraisal of sustainability impacts can be used as basis of understanding the impacts of biofuel projects (collapsed in this case), and what factors must be considered in future project planning to improve their sustainability.

The remainder of this chapter is organized as follows. Section 2 outlines how community participation is conceptualized in biofuel project planning. Section 3 describes the methodology used for the empirical analysis. Section 4 reports the results of focus group discussions and household surveys of local communities at the three collapsed jatropha projects studied. Section 5 discusses the main findings, including the mismatch between the impacts covered in Environmental Impact Assessments (EIAs) reports and those experienced by local communities (Sect. 5.1–5.2), as well as communicating to sustainability science scholars lessons learnt by the authors on how to employ a rapid appraisal of sustainability impacts using the perceptions of local communities (Sect. 5.3).

2 Conceptualizing Community Participation in Biofuel Projects

2.1 Renewable Energy and Community Participation

Across SSA in general, and Ghana in particular, the decentralized planning of renewable energy projects is limited (Ahmed et al. 2014). In developing countries, some decentralized planning systems exist, but they are mostly related to rural-urban planning, rather than energy issues that are mostly tackled at the national level (Hiremath et al. 2007).

In renewable energy contexts, social and community acceptance has three levels of multi-dimensional interactions: (i) community, (ii) market and (iii) socio-political acceptance (Hammami et al. 2016; Wüstenhagen et al. 2007) (Fig. 1). Among these, community acceptance entails mostly issues of trust, procedural

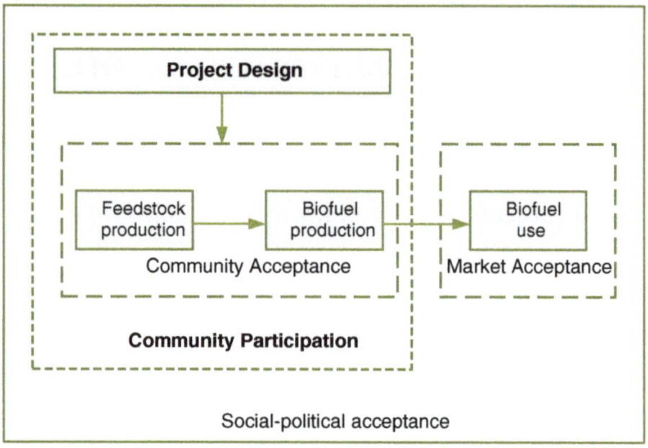

Fig. 1 Conceptual framework of community participation and biofuel acceptance in biofuel commodity chains. *Source* Adapted from (Chin et al. 2014; Wüstenhagen et al. 2007)

and redistributive justice (Hammami et al. 2016; Wüstenhagen et al. 2007). Some scholars make the case that, for biofuel projects to be successfully implemented, the concept of community participation and acceptance must be inextricably linked (Chin et al. 2014). The idea is that community participation during project design is a pre-condition for the biofuel investment to gain community acceptance, build trust between local communities and investors, and achieve procedural and redistributive justice.

However, the conceptualization of "community" (as related to renewable energy projects) has recently been shifting from a territory-based definition (i.e. community as a geographic entity), to wider communities that are not territorially bound and encompass actors who share common values/interests (Hammami et al. 2016). In biofuel commodity chains, these can be seen as 'communities of interest' who have a stake in the biofuel project. The participation of both territory-based and not territory-bounded communities in the design of biofuel project/policies can have significant impacts on their viability and long-term sustainability (e.g. Schut et al. 2014; Slingerland and Schut 2014).

Based on the above the authors view community participation during biofuel project design as an integral element of the social-political acceptance of biofuel projects (Fig. 1). At the community level, project design generally focuses on feedstock and biofuel production (Fig. 1). The link between project design and community acceptance for jatropha is still not well established, but some scholars have hinted that the lack of community acceptance could be linked to limited participation in project design (e.g. Hashim 2014). In renewable energy contexts, it has been argued that appropriate and active participation requires early involvement of communities in the planning process (Polatidis and Haralambopoulos 2004). In biofuel contexts, this means that community views/values should be integrated into the project design before the start of feedstock cultivation (Fig. 1).

2.2 Community Participation in Biofuel Projects in Ghana

During the biofuel boom of the mid-2000s, Ghana (like most SSA countries) did not have the institutional capacity and experience to negotiate land ideas and assess biofuel project designs (and their viability/sustainability) (German et al. 2013). With the inherent bureaucratic nature of decentralization in Ghana, foreign investors were not willing to spare time and financial resources to solicit community views during project design (Boamah 2010, 2011, 2014a, b). In fact, foreign investors mostly planned biofuel projects while awaiting (or often taking for granted) community acceptance (Nyari 2008). However, the government of Ghana eventually put in place a number of policies to regulate the biofuel sector (Ahmed et al. 2015).

The current Bioenergy Act of 2011 indicates the various milestones and requirements that a biofuel company must meet during project planning (Table 1). To reach each milestone the company/investor is required to go through a series of processes, some of which require community participation.

Table 1 Biofuel project planning in Ghana and form of community participation

Legislation milestones	Planning stages	Form of community participation
Acquisition of provisional license	Preparation of business plan	None
	Preparation of financial plan	None
	Implementation plan	None
	Monitoring and evaluation plan	None
Acquisition of siting clearance (Siting permit)	Land conveyance agreement	Land compensation agreement provided by community (*site selection and compensation evaluation*)
	Environmental impact assessment (EIA)	Identify potentially significant impacts of the project that need mitigation measures (*community hearing*)
	Health, safety and environmental plan	Assist biofuel company in providing context-specific appropriate measures (*joint mitigation planning*)
Acquisition of construction work permit (Authorization to construct)	Implementation agreements	Community and company agree on *benefit sharing arrangements*
	Labour agreements	
Acquisition of operational license (Authorization to operate)	Operations and maintenance plan	Community provide final concerns (*site clearing and operation*)
	Safety and technical management plan	None
	Commissioning report	Community attends project commissioning. Development of company-community interface for feedback (*feedback mechanism*)
	Decommissioning plan	Transfer of lands rights, ecological restoration or sustainability of project investments (*Exit planning*)

To get the initial provisional license the company/investor must prepare business, implementation, monitoring and financial plans, that do not require community participation. As most biofuel projects were foreign-led (Schoneveld 2014; Ahmed et al. 2015), the initial business, implementation, monitoring, and financial planning were largely out of community reach. However, Polatidis and Haralambopoulos (2004) argue that early community involvement in such early stages can help build social trust and improve wider acceptance of renewable energy projects.

Community involvement in biofuel project planning commonly starts with site selection as investors move into rural areas in search for appropriate locations to undertake feedstock production. After identifying a potential site, the company/investor has to negotiate with the local community (e.g. chiefs, landowners) and obtain the Land Conveyance Agreement by agreeing on a compensation package. Subsequently, the investor is required to undertake an Environmental Impact Assessment (EIA) for the project, during which community participation takes place through community hearing of the EIA scoping report. At this stage, the community is given the chance to express concerns that need to be addressed by the investor through a mitigation plan that is expected to be planned jointly.

To acquire the Authorization to Construct Permit, the company/investor is expected to provide implementation and labour agreements, which entail negotiations with the community on the format of that benefit sharing agreement (which is then incorporated into the documents). Finally the company/investor is expected to obtain the final license to operate, by providing detailed operational, commissioning and decommission plans. Community participation falls under the purview of deciding the exit plan and the development of company-community interface for feedback during the operation of the project.

3 Methodology

3.1 Study Approach

This chapter undertakes an ex-post rapid appraisal (Sect. 3.3) of the sustainability impacts of three collapsed jatropha projects in Ghana (Sect. 3.2), based on local community perceptions. Our approach draws from sustainability science, which is an ideal framework to study the multiple impacts of biofuels as experienced by different stakeholders (Gasparatos et al. 2013b). This is because sustainability science adopts a use-oriented perspective, connecting social and ecological systems, through a transdisciplinary approach that considers multiple stakeholders (Clark and Dickson 2003; Kates et al. 2007; Kates 2011; Komiyama and Takeuchi 2006).

Even though sustainability science has emerged as a critical research and problem-solving domain, there are still ongoing epistemological discussions regarding its ability to evaluate the 'sustainability' of alternative pathways across different spatial and temporal scales (Kates 2011). For example, it has been suggested that existing tools are not able to comprehensively assess the sustainability of bioenergy projects, as they tend to often adopt a reductionist approach, which often ignores the normative aspects of the bioenergy systems (Buytaert et al. 2012). Reductionist approaches to the assessment of the sustainability impacts of bioenergy systems also leave limited room for stakeholder engagement (Buchholz et al. 2009).

In this chapter the authors attempt to overcome this limitation by adopting a methodology that considers the perceptions of local communities as a means of identifying and assessing key sustainability impacts (Sect. 3.3) of three collapsed jatropha projects in Ghana (Sect. 3.2). The authors draw insights from two major aspects of social acceptance that enable community participation; (i) procedural justice, (ii) redistributive justice. It is inferred from the existing literature that with active community participation (Sect. 2), both procedural and redistributive justice can be met in biofuel project design, which can ultimately have positive implications for the viability and sustainability of biofuel projects (German et al. 2013).

As the Environmental Impact Assessment (EIA) remains maybe the single best opportunity to capture and integrate the views of local communities during the early planning stages of biofuel projects, we juxtapose the impacts covered in the EIAs of three collapsed jatropha projects in Ghana and the impacts eventually experienced by local communities, as captured through our rapid appraisal (Sect. 5.1). We focus on collapsed jatropha projects because they offer an opportunity to understand the impacts of unsustainable biofuel projects, and thus help identify critical factors necessary for designing viable and sustainable biofuel projects in the future. In this respect, the lessons distilled from collapsed jatropha projects can serve as a basis to inform the current and future pathways for biofuel expansion in Ghana (Sect. 5.2).

Rapid appraisal of sustainability impacts has been used in numerous studies of bioenergy systems (Mohr and Raman 2013; Raman et al. 2015a, b; Romijn et al. 2014; Upham and Smith 2013; von Maltitz et al. 2016). They offer several advantages to sustainability science researchers, as they present a quick and simple way to understand broad sustainability patterns over time. By quickly identifying such patterns of change/impact they save significant time and resources when compared to more traditional approaches of capturing community perceptions, such as extensive household surveys combined with geospatial analysis.

In the context of collapsed biofuel projects, such rapid appraisals are possibly the only feasible analytical option. This is because capturing household's exact behavior (e.g. livelihoods, food security, etc.) through extended household surveys becomes highly uncertain if the recollection period exceeds 3–5 years in the past (von Maltitz et al. 2016).

3.2 Study Areas

For the purpose of this chapter, the authors focused on jatropha projects because they represent (i) the major feedstock option promoted by the government of Ghana; (ii) the highest (by far) number of FDIs and large-scale land acquisitions, compared to other feedstocks (e.g. oil palm, sugarcane). As explained earlier, the study identified three collapsed jatropha projects, which represent FDIs from Canada, Norway, and Israel. These projects are among the 20 failed jatropha projects in Ghana that were identified in an earlier scoping study of biofuel projects in the country (Ahmed et al. 2015). Each of the target projects is located within a different ecological zone (guinea savannah, transitional forest, deciduous forest) in order to allow for a comparative study within different agro-ecological settings (Fig. 2, Table 2). The acquired areas ranged between 10,696 and 100,000 ha, but the actual land that was eventually cultivated was much lower (325–1,050 ha) as all projects collapsed in 2011–2012 (within five years of operation).

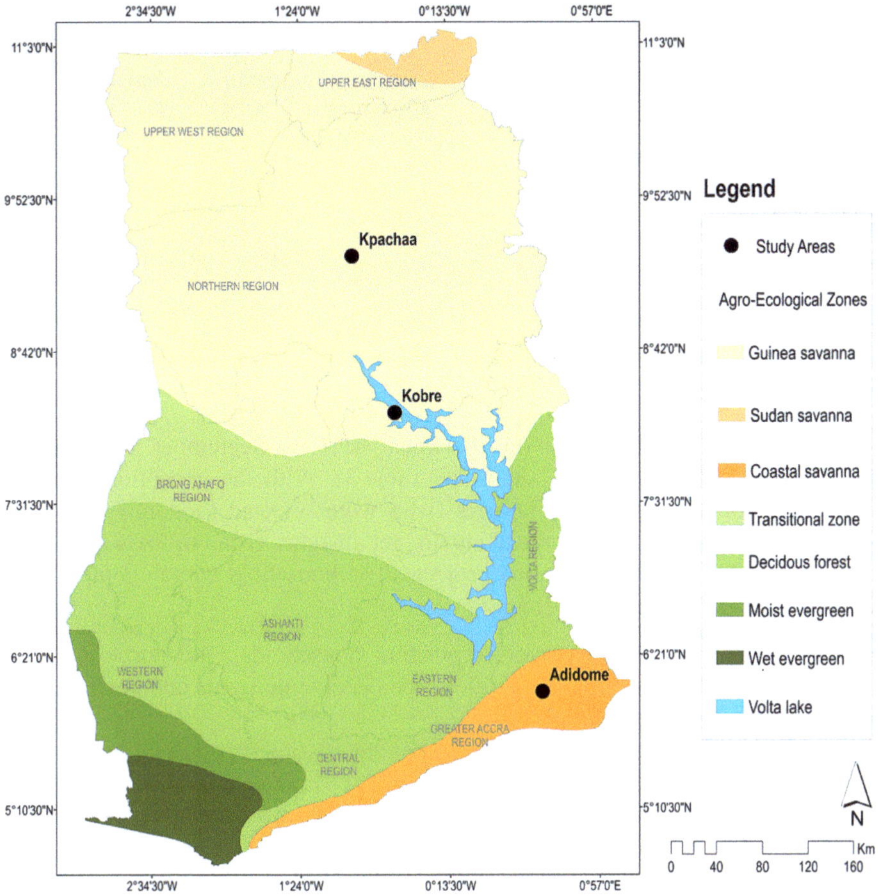

Fig. 2 Map of study sites and other collapsed jatropha projects in Ghana

Table 2 Selection of case study areas (large-scale cultivation)

Location	Project name	Year started	Year collapsed	Land acquired	Land cultivated	Ecosystem type
Kpachaa (Northern region)	BioFuel Africa (Norway)	2006	2011	10,696	1,000	Guinea savannah
Kobre (Brong-Ahafo region)	Kimminic Estates Ltd (Canada/ Ghana)	2007	2012	50,000	1,050	Transitional forest
Adidome (Volta region)	Galton Agro Ltd (Israel)	2008	2012	100,000	325	Deciduous forest

Table 3 Types of respondent for the household survey

Location	Number of Respondents						Total
	Land owners	Native inhabitant	Migrant/ settler farmer	Local government representative	Farm manager	Former plantation workers	
Kpachaa	8	2	5	1	1	13	30
Kobre	7	7	10	1	0	5	30
Adidome	2	15	14	1	1	7	40
Total	17	24	29	3	2	25	100

3.3 Data Collection and Analysis

The authors administered a survey (between the 1–21 September 2015) of 100 households in the three study sites (see Table 3). With the exception of native inhabitants and migrant/settler farmers (who were selected randomly), all other respondents were selected purposively to get insights from those with unique experience with the failed jatropha projects. The household survey comprised of three major parts, as detailed below:

- Part I: demographic aspects of respondent households. The purpose was to understand the household composition, socioeconomic status and gender differentiation in the responses.
- Part II: land acquisition procedures and benefit-sharing (procedural justice component of community acceptance). The purpose was to capture the perspectives of landowners regarding issues that should be incorporated into biofuel project design (e.g. land acquisition processes, compensation and benefit sharing arrangements).
- Part III: sustainability impacts (distributive justice component of community acceptance). The purpose was to undertake a rapid appraisal of the impacts experienced by the local communities, and the distribution of benefits and burdens after collapse. In total 16 sustainability impacts were explored that cut across environmental, social, economic, and cultural impact domains. Data was collected on the community perceptions of the impacts experienced before, during and after jatropha project implementation.

Table 4 highlights the ranking scale adopted for Part III of the household survey. Higher ranking suggests an increasing rate of change for a specific impact (whether positive or negative). The percentage of respondents choosing a score is denoted with a different colour. Essentially, the deeper the shade, the higher the percentage of responses about the direction of impact (implying a stronger consensus between respondents).

Score	Direction	Percentage of Respondents (%)		
		≥80	79–60	40–59
1=Decreased significantly	↓			
2=Decreased little	↘			
3=Remained the same	→			
4=Increased a little	↗			
5=Increased significantly	↑			

Table 4 Ranking scale for Part III of the household survey

4 Results

4.1 Socioeconomic Profile of Study Areas

The gender composition of the respondents was 70.0 % males and 30.0 % females (Table 5). Heads of households represented 75 % of respondents (9 % were female-headed), while 25.0 % were other household members (mainly spouses). Adidome has a disproportionally higher rate of female-headed households (17.5 %). This falls within the recent observed patterns of increasing numbers of female-headed households in rural agrarian communities in Ghana (FAO 2012a, b). Such a pattern is observed due to the shifting gender roles in agricultural activities and the migration of males to urban centers (FAO 2012a, b; Ghana Statistical Service 2014a, b).

The mean age was relatively low at 39.2 years, with a standard deviation of 11.2, (minimum 20 years and maximum 80 years). The average household size was 5.1 with a standard deviation of 2.8. Kpachaa and Kobre had relatively larger household sizes of 6.6 and 5.6, which are greater than the national average of 4.4 (Ghana Statistical Service 2013). Adidome had an average household size of 4, which is less than the national average of 4.4. Approximately 30 % of the respondents did not have any formal education (females accounted for a disproportionally greater proportion of this, at 44 %), with the numbers of respondents with a formal education declining even more for advanced educational levels. The number of respondents without a formal education in Kpachaa was 70 % (21 respondents). This high rate of illiteracy in Kpachaa, which is located in the Northern Region of Ghana, reaffirms the poor human development in that part of the country (Ghana Statistical Service 2014a, b).

When it comes to employment, only 15.0 % of respondents are employed in wage income earning activities. The overwhelming majority (80.0 %) are involved in subsistence farming and petty trading. Within those that are currently employed (15.0 %), 3 % represent workers overseeing machinery left behind after the

Table 5 Summary of background of household survey respondents

Socioeconomic characteristics	Kobre (Kimminic Estates		Kpachaa (BioFuel Africa)		Adidome (Galton Agro Ltd)		Total	
	Freq.	%	Freq.	%	Freq.	%	Freq.	%
Gender of respondent								
Male	24	80.0	22	73.3	24	60.0	70	70.0
Female	6	20.0	8	26.7	16	40.0	30	30.0
Headship of household								
Male	24	80.0	21	70.0	30	75.0	75	75
Female	6	20.0	9	30.0	10	25.0	25	25
Age of respondent								
<30	7	23.3	13	43.4	2	5.0	22	22.0
31–40	9	30.0	10	33.3	19	47.5	38	38.0
41–50	4	13.3	6	20.0	16	40.0	26	26.0
51–60	5	16.7	1	3.3	3	7.5	9	9.0
>60	5	16.7	0	0	0	0.0	5	5.0
Marital status								
Single	7	23.3	1	3.3	5	12.5	13	13.0
Married	22	73.3	29	96.7	28	70.0	79	79.0
Divorced	1	3.3	0	0	7	17.5	8	8.0
Household size								
1–3	4	13.3	5	16.7	21	52.5	30	30.0
4–6	11	36.7	21	70.0	17	42.5	49	49.0
7–9	10	33.3	3	10.0	2	5.0	15	15.0
10–12	2	6.7	1	3.3	0	0.0	3	3.0
>12	3	10.0	0	–	0	0.0	3	3.0
Level of education								
No formal education	7	23.3	21	70.0	2	5.0	30	30.0
Adult literacy	8	26.6	5	16.6	0	0.0	13	13.0
Some basic	2	6.7	0	0.0	17	42.5	19	19.0
Complete basic	2	6.7	0	0.0	11	27.5	13	13.0
Some secondary	2	6.7	0	0.0	3	7.5	5	5.0
Completed secondary	5	16.7	2	6.7	1	2.5	8	8.0
Completed tertiary	4	13.3	2	6.7	6	15.0	12	12.0
Religion								
Christianity	22	73.3	0	0.0	31	77.5	53	53.0
Islam	3	10.0	30	10.0	1	2.5	34	34.0
Traditional	5	16.7	0	0.0	8	20.0	13	13.0
Employment status								
Unemployed	1	3.3	0	0.0	2	5.0	3	3.0
Permanent employ	6	20.0	2	6.7	7	17.5	15	15.0
Self-employment	23	76.7	27	90.0	30	80.0	80	80.0
Retired	0	0.0	0	–	1	2.5	1	1.0

(continued)

Table 5 (continued)

Socioeconomic characteristics	Kobre (Kimminic Estates		Kpachaa (BioFuel Africa)		Adidome (Galton Agro Ltd)		Total	
	Freq.	%	Freq.	%	Freq.	%	Freq.	%
Occupation								
Plantation worker	1	3.3	0	0.0	2	5.0	3	3.0
Civil servant	5	16.7	2	6.7	5	12.5	12	12.0
Student	1	3.3	1	3.3	2	5.0	4	4.0
Self-employment (mainly	23	76.7	27	90.0	30	75.0	80	80.0
subsistence farming)[a]	0	0.0	0	0.0	1	2.5	1	1.0
Retired	0	0.0	0	0.0	1	2.5	1	1.0
Yearly house income (GH¢)								
<1000	1	3.3	9	30.0	5	12.5	15	15.0
1001–5000	17	56.7	17	56.7	30	75.0	65	65.0
5001–10,000	6	20.0	3	10.0	2	5.0	11	11.0
10,001–15,000	4	13.4	0	0.0	3	7.5	7	7.0
15,000–20,000	1	3.3	0	0.0	0	0.0	1	1.0
>20,000	1	3.3	1	3.3	0	0.0	2	2.0

Note [a]For the purposed of this study, family/subsistence farming was classified as an employment activity. However, agricultural produce consumed within the household is not converted into monetary units unless sold to external markets

collapse of jatropha plantations and 12.0 % are civil servants. Household yearly income ranges between GH¢450 (US$118.4)–GH¢25,000 (US$6,578.9), with a mean income of GH¢4,056 (US$1,067.4) and standard deviation of GH¢4,431.3. Many households (65.0 %) earn a yearly income between GH¢1,001–5,000 (UD$263–1,315.8), which falls within the ranges of average annual income in rural areas of Ghana (Ghana Statistical Service 2014a, b).

The above data indicates that the areas where the collapsed jatropha projects are situated are rather poor, with little formal employment opportunities. This means that income from employment at jatropha plantations could have had a discernible effect, even if it represented the lowest minimum agricultural salary in Ghana (GH¢5,100/US$1,345 for formally employed workers in agriculture) (Ghana Statistical Service 2014a). In fact, studies in Ghana have suggested that poor plantation workers appreciate even such low incomes due to its stability (Schoeneveld et al. 2011).

4.2 Community Perceptions of Land Acquisition Processes

Local communities around biofuel projects often hope that they can negotiate with investors or pass on their concerns in order to receive fair benefits (possibly in the form of employment to compensate for the loss of land) and share burdens (Boamah

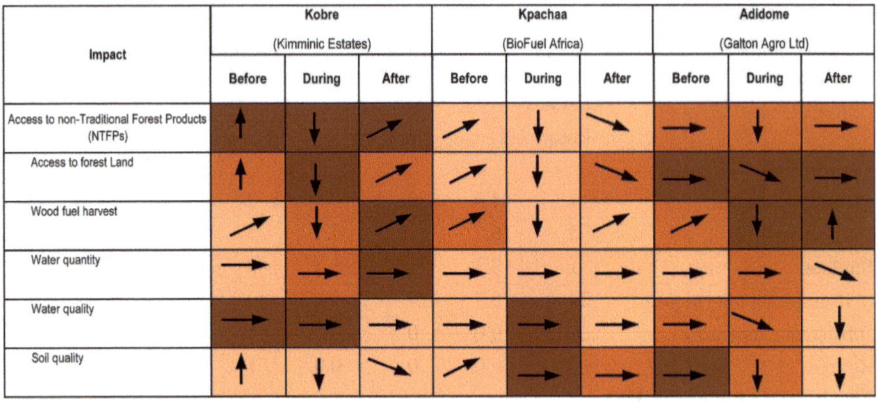

Fig. 3 Environmental impacts experienced in the three study areas

2014b). In Kobre and Kpachaa, jatropha companies/investors acquired land through the local chiefs, while the land in Adidome was acquired from families.

There have been cases of large-scale land acquisitions in Ghana, where the chiefs neither disclosed the amount of transferred land, nor represented their people fairly (Boamah 2014b; Campion and Acheampong 2014). Expert interviews and local surveys showed that this was the case in Kpachaa. Out of the 30 respondents (Table 3), 17 were landowners who gave out an average of 20 acres (8 ha) with no written agreement or contract about benefit sharing/compensation. In fact, only very few respondents (3 out of 30 in Kobre and 2 out of 30 in Kpachaa) received compensation, in terms of either employment or cash. In both areas, the landowners were compelled to give out the land because of instructions from the chiefs. Conversely, in Adidome, families gave out the land because of prospects of better jobs, income, and compensation.

Though our focus is not to judge the adequacy of the compensation packages, potential investors may consider having guidelines for adequate compensation, as a number of jatropha projects collapsed because agitation started due to unfair (or complete lack of) compensation (ActionAid Ghana 2012; Boamah 2011; German et al. 2013). It is worth noting that the lands given for jatropha production were mostly fallow lands.[3] Interestingly, those respondents who reported the loss of fallow land also reported decreases in food availability and agricultural land (see Sect. 4.3, Fig. 3).

It becomes imperative for investors to therefore understand the kinds of land they acquire from communities and to conduct proper consultations. This is because communities might react negatively if their expectations are not met when the loss of land can directly affect their livelihoods. It should be noted at this point that the average fallow land given (8 ha) represents 4 times the average land

[3]Fallow land is any arable land that is generally being cultivated, but is left uncultivated for at least one cropping season in order for it to regain its fertility. Large portions of land concessions acquired by jatropha companies in Ghana were fallow lands (see German et al. 2010).

holding for smallholder or subsistence farmers in Ghana (Ministry of Food and Agriculture 2013).

4.3 Sustainability Impacts Experienced by Local Communities

When it comes to environmental impacts, access to Non-Traditional Forest Products (NTFPs), and access to forests are some of the types of impacts that were felt most severely by the three local communities. Across all case studies, respondents reported a significant decline in access to NTFP during the time of operation of the jatropha companies, though the magnitude of the decline varied among communities (Fig. 3). However, after the collapse of the three projects there were mixed responses in recovery rates. Communities in Kobre reported an increase in access, Kpachaa a slight decrease and Adidome no changes. Given the importance of forests (and their products) for the livelihoods of local communities, such impacts need to be considered during the planning of jatropha projects. This perspective reflects some jatropha-related deforestation impacts raised in literature on Ghana (Boamah 2011; Schoneveld et al. 2011).

For other environmental impacts such as water quality and quantity, there is a unanimous perception across the study sites that jatropha cultivation has not changed the status quo, as all plantations were rain-fed and were not close to water bodies. However, with regards to soil quality there were mixed responses, with Kobre and Adidome reporting declines during the operation of the projects. Kpachaa and Adidome reported a decrease in soil quality after the collapse of the companies, with one former plantation worker in Adidome noting: "I learned the jatropha plant is toxic to the soil and can affect our crop production, but the company did not tell our people the truth".

While there is no scientific evidence to support the above claim,[4] the fact remains that such beliefs hint at the lack of information that the local community had received about the jatropha (and its positive and negative effects). In terms of socioeconomic impacts (Fig. 4), jobs and income generation increased significantly in all communities during the operation of the plantations. Actually, more than 80.0 % of the respondents in each community indicated a significant increase in income during the time of the operation of the project.

It is interesting to note that the income/employment benefits were felt differently amongst the various households. In particular, across all study sites low-income households who earn below the annual average income of GH¢4,056 (US$1,067.4) reported having experienced much more significant effects on jobs and income. For example, out of 33 respondents who reported a significant increase in income, 54.5 % came from households who earn less than the

[4]Jatropha is indeed toxic if consumed by humans and animals (Devappa et al. 2010), but as a tree species its roots could potentially bind the soil, possibly reducing soil erosion (Openshwa 2000).

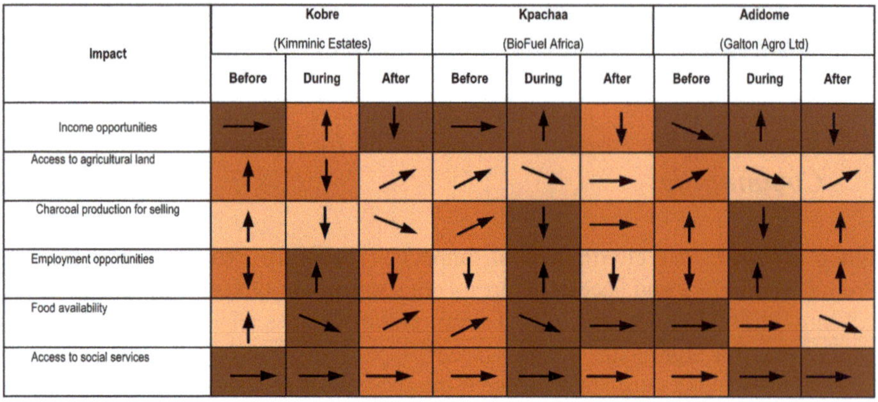

Fig. 4 Socioeconomic impacts experienced in the three study areas

Ghanaian annual average income. Also, out of the 34 respondents who reported a significant increase in jobs available during project implementation, 55.9 % earned below the annual average income in Ghana. However, after the collapse of the companies, jobs and income opportunities declined rapidly in all communities, as there were no other significant sources of rural employment in the study areas. This suggests that the employment and income opportunities created during the operation of the plantations could have had some local poverty alleviation effects, though these were lost after the project collapsed.

However, it is important to point out some interesting trade-offs between the different socioeconomic impacts. For example, while job/income opportunities increased, other traditional sources of livelihoods such as charcoal-making experienced a decline (Fig. 4). This is not surprising given the lack of access to forest resources, as discussed above (Fig. 3). Food availability is also perceived to have declined during the operation of the jatropha projects, possibly due to the loss of fallow land (Sect. 4.2) and restriction in picking wild fruits in the land owned by companies. However, after the projects collapsed several households encroached the lands of the former plantations to conduct subsistence cultivation, thereby increasing food availability (according to the perspective of respondents, see Fig. 4).

Access to social services (e.g. boreholes, schools, clinics) does not seem to have changed over time. This suggests that the jatropha companies did not develop any social services for the benefit of the communities during their operation, despite some plans to develop such services. For example, the EIA report of BioFuel Africa in Kpachaa indicates plans for the development of school scholarship and borehole water services for the community (Centre for Environment and Health Research and Training 2010a). However, at the time the authors visited the site (September 2015) these services were not being provided, even though BioFuel Africa operated for some 4 years before it collapsed.

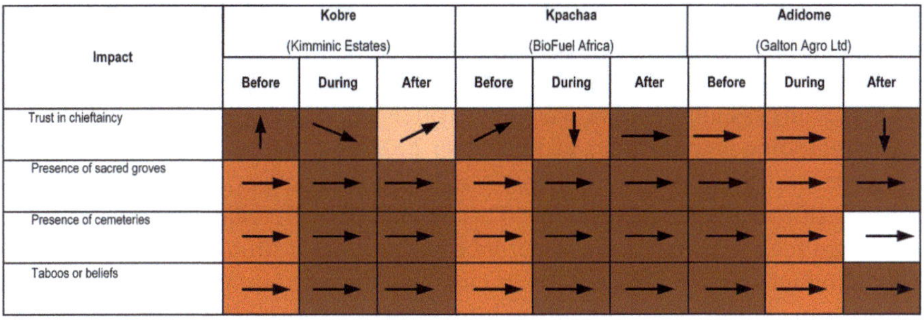

Fig. 5 Cultural impacts experienced in the three study areas

The potential cultural impacts of biofuel production have received some attention in biofuel literature (Boamah 2014b; Campion and Acheampong 2014; Timko et al. 2014). In the communities the authors visited, respondents generally did not perceive any negative effect of the plantations on sacred groves and local taboos (Fig. 5). However, it is worth noting a decline in the trust in chieftaincy institutions, which reflects the findings of other studies in Ghana (Boamah 2014a, b; Campion and Acheampong 2014). The lack of transparency and accountability by chiefs has served as a driver of collapse in some jatropha projects in Ghana (Campion and Acheampong 2014). The decline in trust to local chiefs further indicates the need for potential investors to widen up consultation processes, especially maintaining transparency and accountability between chiefs, the company, and local communities.

4.4 Gender-Differentiation of Experienced Impacts

Though there is a significant gap between male and female respondents (see Table 5), for some impact categories the response patterns are skewed according to gender. For example, in Kpachaa out of 14 respondents reporting a significant decrease in access to non-traditional forest products (NTFPs), 64.3 % were females. This could be a result of the dominance of women in the collection of NTFPs, such as Shea fruits for the Shea butter industry (Collins et al. 2014). Similarly, in Adidome no male respondents indicated a significant decrease in fuelwood harvest (out of 40 respondents). Also, out of 10 people in the community who reported a significant decrease in agricultural land 6 were females, as part of the land used for the plantation was an area where women used to cultivate vegetables such as pepper. These trends reflect well the agrarian development literature in rural Ghana, where both the picking of wild fruits and the collection of fuelwood are considered as female activities (Carr 2008; Elias and Carney 2007; Pouliot 2012).

On the other hand, all of the 17 landowners who provided land to the companies were males. This also agrees with existing development narratives in Ghana and other parts of the global south where gender scholarship has the ubiquitous (if not universal) assumption that women are often neglected in decision-making processes regarding land (Damisa and Yohanna 2007; Rocheleau et al. 1996, 2006).

The results above demonstrate that some of the impacts experienced might have been felt differently by each gender. This means that different parts of the community might feel certain impacts of jatropha projects more strongly, as a reflection of their different roles in the agrarian communities where jatropha production takes place. As a result, it is advisable that consultative processes during the design of biofuel projects capture the perspectives of both male and female community members.

5 Discussion

5.1 Comparing the Impacts Captured in EIA Reports and Experienced by Local Communities

The EIA process remains maybe the single best opportunity for communities in Ghana to voice (and integrate) their concerns during the early planning stages of biofuel projects. However, the comparison of the results of our household survey/expert interviews (Sect. 4.3) and the EIAs for the three collapsed projects suggests mismatches between the impacts covered in the EIAs and those perceived/experienced by local communities (Table 6).

While the EIAs covered a comprehensive list of potential environmental impacts, each of the local communities experienced only some of these impacts. In fact, the environmental impacts perceived by local communities tended to gravitate towards issues related directly to their livelihoods, such as the loss of forestland and access to NTFPs/fuelwood.[5] However, it is worth noting that environmental impacts related to water availability/quality and soil quality were not explicitly raised in our household surveys (Table 6).[6]

Conversely, socioeconomic impacts were felt much more strongly by local communities. In particular, impacts related to employment/income generation, loss of traditional livelihoods and food availability received the highest scores by most respondents. However, the extent to which the EIAs reviewed integrated such impacts was limited (Centre for Environment and Health Research and Training 2010a, b). This reflects criticisms that current EIA approaches tend to have narrow scopes (Jay et al. 2007; Lenzen et al. 2003).

[5]Such points have also been made in other contexts of industrial crop expansion such as oil palm in Indonesia (Moreno-Peñaranda et al. 2015).

[6]Air pollution and GHG emissions were not considered in our study as they either manifest at others areas (e.g. sites of biofuel combustion) or do not have explicitly local effects.

Table 6 Sustainability impacts considered in EIAs and experienced by local communities

Impact	EIA reports						Impacts reported by community		
	Impacts considered			Mitigation plans developed					
	Kobre (Kimminic Estates	Kpachaa (BioFuel Africa)	Adidome (Galton Agro Ltd)	Kobre (Kimminic Estates)	Kpachaa (BioFuel Africa)	Adidome (Galton Agro Ltd)	Kobre (Kimminic Estates)	Kpachaa (BioFuel Africa)	Adidome (Galton Agro Ltd)
Environmental									
Access to Non-Timber Forest Products (NTFPs)	✓	✓	✓	X	✓	✓	✓	✓	✓
Access to forestland	✓	✓	✓	✓	✓	✓	✓	✓	✓
Wood fuel harvest	X	X	X	X	X	X	✓	✓	✓
Water quantity	✓	✓	✓	✓	✓	✓	?	?	?
Water quality	✓	✓	✓	✓	✓	✓	?	?	?
Soil quality	✓	✓	✓	✓	✓	✓	?	?	?
Air pollution	✓	✓	✓	X	X	✓	X	X	X
GHG emissions	X	✓	✓	X	X	✓	X	X	X
Waste generation	X	✓	✓	X	X	✓	X	X	X
Socioeconomic									
Income opportunities	✓	✓	✓	–	–	–	✓	✓	✓
Access to agricultural land	✓	✓	✓	X	X	X	✓	✓	✓
Charcoal production for selling	X	X	X	X	X	X	✓	✓	✓
Job opportunities	✓	✓	✓	–	–	–	✓	✓	✓
Food availability	X	X	X	✓	✓	X	✓	✓	✓
Access to social services	✓	✓	✓	✓	✓	✓	?	?	?

(continued)

Table 6 (continued)

Impact	EIA reports						Impacts reported by community		
	Impacts considered			Mitigation plans developed					
	Kobre (Kimminic Estates	Kpachaa (BioFuel Africa)	Adidome (Galton Agro Ltd)	Kobre (Kimminic Estates)	Kpachaa (BioFuel Africa)	Adidome (Galton Agro Ltd)	Kobre (Kimminic Estates)	Kpachaa (BioFuel Africa)	Adidome (Galton Agro Ltd)
Cultural									
Trust in chieftaincy	X	X	X	X	X	X	✓	✓	?
Presence of sacred groves	X	X	X	X	X	X	?	?	?
Presence of cemeteries	X	X	X	X	X	X	?	?	?
Taboos or beliefs	✓	✓	X	✓	✓	X	?	?	?

Note ✓ Impact/mitigation strategy considered in EIA or impact experienced by community (as captured through the rapid sustainability appraisal)
X Impact/mitigation strategy not considered in EIA or impact not experienced by community (as captured through the rapid sustainability appraisal)
? Impact experienced by a relatively modest fraction of respondents (typically articulated by less than 50 % of respondents during the rapid sustainability appraisal)
– No mitigation measures can be developed as these represent positive impacts
Sources Adanes Consult (2011), Centre for Environment and Health Research and Training (2010a, b) and authors' own research

This mismatch between the impacts covered by EIAs and those experienced by local communities might have ramifications for the proper consideration of community concerns during project planning, especially when it comes to the impacts experienced. In this sense, the overemphasizing or downplaying of certain impacts suggests that EIA processes in the three projects were a missed opportunity for seriously integrating community concerns into project design.

This is also partly reflected in the selection of mitigation measures. Though some of the mitigation measures proposed in the EIAs reflect some of the impacts that strongly felt by communities, other very important mitigation measures were actually not considered, e.g. measures to manage the loss of access to the agricultural land transferred to companies. Furthermore, in reality not all of the proposed mitigation measures were implemented because the EIA reports were only approved in 2010 (for Kobre and Kpachaa) and 2011 (for Adidome), shortly before the collapse of all projects. This long lag between the start of the projects' operation and EIA completion suggests that EIAs were most likely conducted to meet legal obligations, rather than because biofuel companies were honestly concerned to minimize the negative potential impacts of their operations on local communities.

5.2 Key Observations for Sustainability

Our results highlight the emergence of sustainability trade-offs due to the operation and subsequent collapse of jatropha projects in Ghana. Local community perceptions suggest that, while biofuel projects generated employment and income in poor rural settings, they became detrimental to other traditional livelihood activities such as charcoal making (due to loss of access to forest areas) and subsistence food production (due to transfer of fallow lands). Minimizing such trade-offs is a central challenge for enhancing the sustainability of future biofuel production in Ghana and other parts of Africa. Whichever mode of feedstock production is considered (be it smallholder-based, plantation-based or hybrid), capturing community perceptions early on can offer good insights into context-specific realities necessary for identifying and minimizing such sustainability trade-offs.

A second observation has to do with the lack of third-party certification in all three jatropha plantations. None were certified by the Roundtable on Sustainable Biomaterials (RSB), or any another appropriate body.[7] Our results show that the jatropha project in Kpachaa faced instances of land dispossession and land disputes, while the projects in Adiome and Kobre had negative environmental impacts. These are all key principles enshrined in the RSB Certification criteria (RSB 2013). Meyer and Priess (2014) suggest that certification could enhance

[7]To our best knowledge the only biofuel projects certified by RSB in Sub-Sahara Africa are Addax Bioenergy (Sierra Leone) and Solaris (South Africa) (RSB: personal communication).

biofuel sustainability, as it is more likely to influence the adoption of sustainability principles and good practices, during the design and implementation phases of biofuel projects (Dale and Ong 2014). Third-party certification could have also provided an opportunity for the projects studied to be improved from the start, and possibly would have helped improve their overall chances of success.

A third observation that can be made relates to the power of capturing community perceptions as a means of identifying the sustainability impacts of biofuel projects during the different project phases. While some studies have developed different sustainability criteria for the assessment of bioenergy projects (e.g. Acosta et al. 2013, 2014; Acosta-Michlik et al. 2011; Meyer and Priess 2014), they often adopt a top-down approach (i.e. pre-determined criteria through expert judgment) (Sect. 5.3). Our study makes the case that capturing such perceptions early on could inform the selection of sustainability impacts to be considered in EIAs, and potentially improve the participation of local communities during the planning of biofuel projects (Sect. 5.3).

5.3 Implications for Sustainability Science Studies

When it comes to bioenergy systems, several studies have advocated the need to adopt a unified sustainability appraisal approach to study their impacts (Gasparatos et al. 2013b; Liew et al. 2014; Mohr and Raman 2013; Boucher et al. 2014; Suwelack and Wüst 2015). However, sometimes such unified approaches assume, both implicitly and explicitly, the values of affected stakeholders in the criteria and tools used (Gasparatos 2010; Gasparatos and Scolobig 2012). Expert judgment during the development and application of sustainability appraisal approaches can widen the gap between society and experts, and reduce the extent to which local communities can meaningfully participate when determining the sustainability impacts that should be assessed. Considering the high context-specificity of the ecological/social settings of feedstock production, community values and biofuel policies, adopting a unified sustainability appraisal approach could undermine the proper consideration of some local issues (Gasparatos et al. 2013b).

When it comes to the assessment of sustainability impacts of biofuel projects, both top-down and bottom-up appraisal approaches can be used (Fig. 6). While top-down approaches can be useful, their criteria and indicators might miss some of the context-specific realities and localized impacts of biofuel projects. This was the case in our study sites, where even with the use of national EIA criteria, we observed a mismatch between the impacts assessed during the EIA and the actual impacts experienced by local communities (Sect. 5.1). This raises the challenge of how to translate global or regional criteria to capture context-specific local realities. We believe that bottom-up approaches could help identify critical impacts (and select relevant indicators) during the sustainability assessment of bioenergy projects.

In our study we employed a rapid sustainability appraisal approach using local community perceptions to understand impacts during the operation and after the

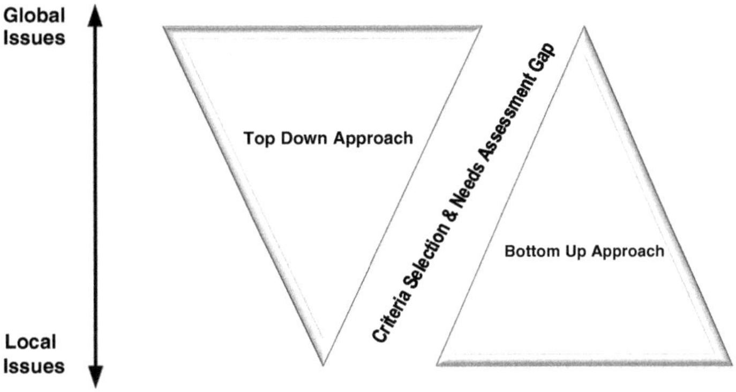

Fig. 6 *Bottom-up* and *top-down* sustainability impacts assessment approaches

collapse of three jatropha projects in Ghana. This is closer to the concept of a bottom-up sustainability impact assessment (Fig. 6). Apart from saving resources and being possibly the only viable alternative to study collapsed jatropha projects (Sect. 3.1), rapid appraisals offer certain advantages. First, they have high explanatory power when capturing local sustainability impacts as they make use of insights provided by local communities that have an intimate knowledge of local contexts as they often experience first-hand the positive/negative impacts of bioenergy projects. Second, if used early in the planning process they can allow for a better alignment of project goals with community needs. This could improve the choice/design of environmental mitigation strategies or social interventions to address the negative impacts expected to arise during the operation of biofuel projects. Such an approach can be undertaken at the onset of the planning phase and be used as a basis not only for the EIA and the development of mitigation strategies, but also for the continuous evaluation of the project.

On the other hand rapid bottom-up sustainability appraisals also have some disadvantages. First of all, soliciting community points of view requires significant time resource (albeit less than more conventional approaches, Sect. 3.1, Khodyakov et al. 2013) that make top-down approaches sometimes very appealing (Mohr and Raman 2013; Boucher et al. 2014). Furthermore, while in our study we could identify impact patterns quickly (Figs. 3, 4 and 5), the exact quantification of trade-offs was not possible. It was also not possible to properly identify the mechanisms that led to the impacts. This suggests that when using rapid appraisals there is a trade-off between the speed of the analysis and the richness of the results obtained.

For sustainability science scholars, rapid appraisals can be used to initially understand the study area and inform how to best frame the actual data collection that will be obtained through subsequent fieldwork. This preliminary scoping step is important to help inform and strengthen the development and design of the methodology and field survey. In this context, rapid appraisals should not be mistaken as a piloting of the actual data collection instruments (e.g. household surveys, focus groups or expert interviews), but rather they form part of the problem identification process before designing the actual methodology.

6 Conclusions

Jatropha production in Ghana was promoted to meet the professed goals of the national biofuel programme, i.e. rural development and energy security (Ahmed et al. 2015). However, the operation and collapse of jatropha projects not only failed to contribute to meeting these objectives, but also had negative impacts on local communities.

In this chapter, the authors identified the impacts experienced by local communities around three collapsed jatropha projects through a rapid sustainability appraisal using community perceptions. Our methodology draws insights from the concepts of social acceptance, and procedural and redistributive justice. The results suggest that certain trade-offs emerged between different sustainability impacts, and highlight the need for community participation during the planning of biofuel projects. In particular, issues related to land acquisition procedures were prevalent in all study sites, as most landowners did not receive any compensation for ceding land (mostly fallow agricultural land) or documentation to prove that the land deal had been carried out. These processes are often described as unjust and create mistrust between the different stakeholders involved in biofuel value chains (German et al. 2013; Nyari 2008).

In all project sites the comparison of the results of the rapid sustainability appraisal and the EIAs reveals an important mismatch between the sustainability impacts included in the EIAs and those actually experienced by local communities. The former had a strong emphasis on some environmental impacts (air, water, soil pollution) and loss of tree species, rather than long-term and indirect environmental effects, which are linked to the socioeconomic circumstances of the community. We argue that this mismatch is the product, to an extent, of the limited (or lack of) active participation of local communities during biofuel project planning and design.

Based on this premise, we propose that a bottom-up sustainability assessment using community perceptions can go a long way towards identifying and selecting relevant criteria during the impact assessment of bioenergy projects (Fig. 6). However, given the inherent uncertainties of this approach, we recommend sustainability science students and practitioners to use such rapid appraisals at the early stages of research. Rapid appraisals can be very powerful techniques to identify patterns that can be used to inform the framing of the research and strengthen the development of data collection methodologies before actual fieldwork takes place.

Finally, as many countries in SSA promote biofuel feedstock production it is important to improve the planning of biofuel projects by enhancing the participation of local communities. The development of stronger company-community interfaces and broadening the scope of EIAs are imperative for the successful participation of local communities during the planning stages of biofuel projects. Company-community interfaces and wider EIA consultation mechanisms can create the preconditions for social acceptance, and procedural and redistributive justice in biofuel commodity chains. We believe that meaningful efforts to involve communities in all stages of the planning cycle of biofuel projects can go a long way towards enhancing the overall viability and sustainability of these projects.

Acknowledgements The authors acknowledge financial support of the Japan Science and Technology Council (JST) through the funding of Belmont Forum project FICESSA. AA is supported by a Monbukagakusho scholarship offered by the Japanese Ministry of Education, Culture, Sports, Science and Technology (MEXT) and through the Graduate Program in Sustainability Science–Global Leadership Initiative (GPSS-GLI), at the University of Tokyo.

References

Achten, W. M. J., Mathijs, E., Verchot, L., Singh, V. P., Aerts, R., & Muys, B. (2007). *Jatropha* biodiesel fueling sustainability? *Biofuels, Bioproducts and Biorefining, 1*, 283–291.

Acosta, L. A., Enano, N. H., Magcale-Macandog, D. B., Engay, K. G., Herrera, M. N. Q., Nicopior, O. B. S., & Lucht, W. (2013). How sustainable is bioenergy production in the Philippines? A conjoint analysis of knowledge and opinions of people with different typologies. *Applied Energy, 102*, 241–253. doi:10.1016/j.apenergy.2012.09.063.

Acosta, L. A., Eugenio, E. A., Enano, N. H., Magcale-Macandog, D. B., Vega, B. A., Macandog, P. B. M., & Lucht, W. (2014). Sustainability trade-offs in bioenergy development in the Philippines: An application of conjoint analysis. *Biomass and Bioenergy, 64*, 20–41. doi:10.1016/j.biombioe.2014.03.015.

Acosta-Michlik, L., Lucht, W., Bondeau, A., & Beringer, T. (2011). Integrated assessment of sustainability trade-offs and pathways for global bioenergy production: Framing a novel hybrid approach. *Renewable and Sustainable Energy Reviews, 15*(6), 2791–2809. doi:10.1016/j.rser.2011.02.011.

ActionAid Ghana. (2012). Land grabbing, biofuel investment and traditional authorities in Ghana: Policy brief. Available at: http://www.actionaid.org/sites/files/actionaid/policy_brief.pdf

Adanes Consult. (2011). *Final environmental impact assessment: Proposed jatropha farm project*. Ghana, Accra: Adidome.

Ahmed, A., Gasparatos, A., & Campion, B. B. (2015). Biofuel development in Ghana: policies of expansion and drivers of failure in the jatropha sector (submitted to Renewable and Sustainable Energy Reviews).

Ahmed, A., Kanton, S., Godwin, K., Rahim, A. A., & Salia, R. A. (2014). Biofuel development and large scale land acquisitions in Ghana, implications for land use planning. *International Journal of Development Research, 4*(12), 2563–2571.

Aitken, M. (2010). Wind power and community benefits: Challenges and opportunities. *Energy Policy, 38*(10), 6066–6075.

Boamah, F. (2010). *The paradox of global discourses and Ghana's biofuel debate: Evidence from biofuel investments in Northern Ghana.* A paper presented at the Agrarian and Rural Development in Africa: Stakeholder's Conference in Harare, Zimbabwe between 24th and 26th November 2010.

Boamah, F. (2011).*The relationship between land grabbing for biofuels and food security, a bane or boon? The food security implications of jatropha biodiesel project in Northern Ghana*. Paper presented at the International Conference on Global Land Grabbing, Future Agricultures Consortium, Sussex.

Boamah, F. (2014a). Imageries of the contested concepts 'land grabbing' and 'land transactions': Implications for biofuels investments in Ghana. *Geoforum, 54*, 324–334.

Boamah, F. (2014b). How and why chiefs formalise land use in recent times: The politics of land dispossession through Biofuels investments in Ghana. *Review of African Political Economy, 41*(141), 406–423.

Boucher, P., Smith, R., & Millar, K. (2014). Biofuels under the spotlight: The state of assessment and potential for integration. *Science and Public Policy, 41*(3), 283–293. doi:10.1093/scipol/scu028.

Buchholz, T., Rametsteiner, E., Volk, T. A., & Luzadis, V. A. (2009). Multi Criteria Analysis for bioenergy systems assessments. *Energy Policy, 37*(2), 484–495. doi:10.1016/j.enpol.2008.09.054.

Campion, B. B., & Acheampong, E. (2014). The Chieftaincy Institution in Ghana: Causers and Arbitrators of Conflicts in Industrial Jatropha Investments. *Sustainability, 6*, 6332–6350.

Carr, E. R. (2008). Men's crops and women's crops: the importance of gender to the understanding of agricultural and development outcomes in Ghana's Central Region. *World Development, 36*(5), 900–915.

Cass, N., Walker, G., & Devine-Wright, P. (2010). Good neighbours, public relations and bribes: The politics and perceptions of community benefit provision in renewable energy development in the UK. *Journal of Environmental Policy & Planning, 12*(3), 255–275.

Centre for Environment and Health Research and Training. (2010a). *Final environmental impact assessment: Proposed jatropha farm project*. Ghana, Accra: Kpachaa.

Centre for Environment and Health Research and Training. (2010b). *Final environmental impact assessment: Proposed jatropha farm project*. Ghana, Accra: Kobre.

Chin, H.-C., Choong, W.-W., Wan Alwi, S. R., & Mohammed, A. H. (2014). Issues of social acceptance on biofuel development. *Journal of Cleaner Production, 71*, 30–39.

Clancy, D., Breen, J., Moran, B., Thorne, F., & Wallance, M. (2011). Examining the socioeconomic factors affecting willingness to adopt bioenergy crops. *Journal of International Farm Management, 5*(4), 25–40.

Clark, W. C., & Dickson, N. M. (2003). Sustainability science: The emerging research program. *Proceedings of the National Academy of Sciences of the United States of America, 100*(14), 8059–8061.

Collins, A. M., Suleman, D., & Ricky, Y. N. (2014). Urban poverty in Northern Ghana: Tracing the livelihood strategies of Women in the Shea Butter Industry. *Journal of Humanity and Social Science Research, 3*, 15–25.

Cowell, R., Bristow, G., & Munday, M. (2011). Acceptance, acceptability and environmental justice: the role of community benefits in wind energy development. *Journal of Environmental Planning and Management, 54*(4), 539–557.

Dale, B. E., & Ong, R. G. (2014). Design, implementation, and evaluation of sustainable bioenergy production systems. *Biofuels, Bioproducts and Biorefining, 8*(4), 487–503. doi:10.1002/bbb.1504.

Damisa, M. A., & Yohanna, M. (2007). Role of rural women in farm management decision making process: ordered probit analysis. *World Journal of Agricultural Sciences, 3*(4), 543–546.

De Schutter, O. (2011a). The green rush: The global race for farmland and the rights of land users. *Harvard International Law Journal, 52*(2), 503–559.

De Schutter, O. (2011b). How not to think of land-grabbing: Three critiques of large-scale investments in farmland. *Journal of Peasant Studies, 38*, 249–279.

Devappa, R. K., Makkar, H. P. S., & Becker, K. (2010). Jatropha toxicity–a review. *Journal of Toxicology and Environmental Health. Part B, Critical Reviews, 13*(6), 476–507. doi:10.1080/10937404.2010.499736

Devine-Wright, P. (2007). Energy citizenship: Psychological aspects of evolution in sustainable energy technologies. In: J. Murphy (Ed), *Framing the Present, Shaping the Future: Contemporary Governance of Sustainable Technologies*. Earthscan, pp. 63–86

Dragojlovic, N., & Einsiedel, E. (2015). What drives public acceptance of second-generation biofuels? Evidence from Canada. *Biomass and Bioenergy, 75*, 201–212.

Elias, M., & Carney, J. A. (2007). African shea butter: A feminized subsidy from Nature, Africa. *Journal of the International African Institute, 77*, 37–62.

FAO. (2012a). *Gender inequalities in rural employment in Ghana: An overview*. Rome, Italy: Gender, Equity and Rural Employment Division of FAO.
FAO. (2012b). *Gender inequalities in rural employment in Ghana: Policy and legislation*. Rome, Italy: Gender, Equity and Rural Employment Division of FAO.
Gasparatos, A. (2010). Embedded value systems in sustainability assessment tools and their implications. *Journal of Environmental Management, 91*(8), 1613–22. doi:10.1016/j.jenvman.2010.03.014.
Gasparatos, A., & Scolobig, A. (2012). Choosing the most appropriate sustainability assessment tool. *Ecological Economics, 80*, 1–7. doi:10.1016/j.ecolecon.2012.05.005.
Gasparatos, A., Lee, L., von Maltitz, G., Mathai, M., Puppim de Oliveira, J., Johnson, F. X., & Willis, K. (2013a). Catalysing biofuel sustainability: International and national policy interventions. *Environmental Policy and Law, 43*, 216–221.
Gasparatos, A., Lehtonen, M., & Stromberg, P. (2013b). Do we need a unified appraisal framework to synthesize biofuel impacts? *Biomass and Bioenergy, 50*, 75–80. doi:10.1016/j.biombioe.2012.09.052.
Gasparatos, A., von Maltitz, G. P., Johnson, F. X., Lee, L. Y., Mathai, M. V., Puppim de Oliveira, J. A., & Willis, K. J. (2015). Biofuels in sub-Sahara Africa: Drivers, impacts and priority policy areas. *Renewable and Sustainable Energy Reviews, 45*, 879–901.
German, L., Schoneveld, G., & Mwang, E. (2013). Contemporary processes of large-scale land acquisition in sub-Saharan Africa: legal deficiency or elite capture of the rule of law. *World Development, 48*, 1–18.
German, L., Schoneveld, G., Skutsch, M, Andriani, R., Obidzinski, K., & Pacheco, P., et al. (2010). The local social and environmental impacts of biofuel feedstock expansion: a synthesis of case studies from Asia, Africa and Latin America. CIFOR Infobrief No.34. Bogor, Indonesia, Center for International Forestry Research (CIFOR).
Ghana Statistical Service. (2013). *2010 Population and housing census: National analytical report*. Accra, Ghana: GSS.
Ghana Statistical Service. (2014a). *Ghana living standards survey round 6: Labour force report*. Accra, Ghana: GSS.
Ghana Statistical Service. (2014b). *Ghana living standards survey round 6: Poverty profile in Ghana*. Accra, Ghana: GSS.
Giannoccaro, G., & Berbel, J. (2012). The determinants of farmer's intended behavior towards the adoption of energy crops in Southern Spain: an application of the classification tree-method. *Bio-based and Applied Economics, 1*(2), 199–212.
Gunatilake, H., Roland-Holst, D., & Sugiyarto, G. (2014). Energy security for India: Biofuels, energy efficiency and food productivity. *Energy Policy, 65*, 761–767.
Hammami, S. M., chtourou, S., & Triki, A. (2016). Identifying the determinants of community acceptance of renewable energy technologies: The case study of a wind energy project from Tunisia. *Renewable and Sustainable Energy Reviews, 54*, 151–160.
Hashim, N. (2014). How knowledge, policy planning, and implementation succeed or fail: The Jatropha projects in Tanzania. *Journal of African Business, 15*(3), 169–183.
Hinshelwood, E. (2001). Power to the people: Community-led wind energy—obstacles and opportunities in a South Wales Valley. *Community Development Journal, 36*(2), 96–110.
Hiremath, R. B., Shikha, S., & Ravindranath, N. H. (2007). Decentralized energy planning; modeling and application—a review. *Renewable and Sustainable Energy Reviews, 11*(5), 729–752.
Jay, S., Jones, C., Slinn, P., & Wood, C. (2007). Environmental impact assessment: Retrospect and prospect. *Environmental Impact Assessment Review, 27*(4), 287–300.
Johnson, F. X., & Silveira, S. (2014). Pioneer countries in the transition to alternative transport fuels: Comparison of ethanol programmes and policies in Brazil, Malawi and Sweden. *Environmental Innovation and Societal Transitions, 11*, 1–24
Kates, R. W. (2011). What kind of science is sustainability science? *Proceedings of the National Academy of Science USA, 108*(48), 19449–19450.

Khodyakov, D., Stockdale, S., Jones, A., Mango, J., Jones, F., & Lizaola, E. (2013). On measuring community participation in research. *Health Education & Behavior: The Official Publication of the Society for Public Health Education, 40*(3), 346–54. doi:10.1177/1090198112459050.

Kidido, J. K., & Kuusaana, E. D. (2014). Large-scale investment in biofuel feedstock production and emerging land issues in Ghana. *Journal of Social Science Studies, 1*(2), 163–176.

Komiyama, H., & Takeuchi, K. (2006). Sustainability science: Building a new discipline. *Sustainability Science, 1*, 1–6.

Kuntashula, E., van der Horst, D., & Vermeylen, S. (2014). A pro-poor biofuel? Household wealth and farmer participation in Jatropha curcas seed production and exchange in eastern Zambia. *Biomass and Bioenergy, 63*, 187–197.

Lenzen, M., Murray, S. A., Korte, B., & Dey, C. J. (2003). Environmental impact assessment including indirect effects—a case study using input–output analysis. *Environmental Impact Assessment Review, 23*(3), 263–282.

Liew, W. H., Hassim, M. H., & Ng, D. K. S. (2014). Review of evolution, technology and sustainability assessments of biofuel production. *Journal of Cleaner Production, 71*, 11–29. doi:10.1016/j.jclepro.2014.01.006.

Longstaff, H., Secko, D. M., Capurro, G., Hanney, P., & McIntyre, T. (2015). Fostering citizen deliberations on the social acceptability of renewable fuels policy: The case of advanced lignocellulosic biofuels in Canada. *Biomass and Bioenergy, 74*, 103–112.

Månsson, A., Sanches-Pereira, A., & Hermann, S. (2014). Biofuels for road transport: Analysing evolving supply chains in Sweden from an energy security perspective. *Applied Energy, 123*, 349–357.

Meyer, M. A., & Priess, J. A. (2014). Indicators of bioenergy-related certification schemes—an analysis of the quality and comprehensiveness for assessing local/regional environmental impacts. *Biomass and Bioenergy, 65*, 151–169. doi:10.1016/j.biombioe.2014.03.041.

Ministry of Food and Agriculture. (2013). Agriculture in Ghana: Facts and figure. Statistics, Research and Information Directorate (SRID), Accra, Ghana.

Mitchell, D. (2011). *Biofuels in Africa*. Washington: TheInternaitonal Bank for Reconstruction and Development/World Bank.

Mohr, A., & Raman, S. (2013). Lessons from first generation biofuels and implications for the sustainability appraisal of second generation biofuels. *Energy Policy, 63*(100), 114–122. doi:10.1016/j.enpol.2013.08.033.

Montefrio, M. J. F., Sonnenfeld, D. A., & Luzadis, V. A. (2015). Social construction of the environment and smallholder farmers' participation in "low-carbon", agro-industrial crop production contracts in the Philippines. *Ecological Economics, 116*, 70–77.

Moreno-Peñaranda, R., Gasparatos, A., Stromberg, P., Suwa, A., Pandyaswargo, A. H., & Puppim de Oliveira, J. A. (2015). Sustainable production and consumption of palm oil in Indonesia: What can stakeholder perceptions offer to the debate? *Sustainable Production and Consumption, 4*, 16–35.

Nyari, B. (2008). *Biofuel land grabbing in Northern Ghana*. Ghana and African Biodiversity Network: Regional Advisory and Information Network Systems.

Openshaw, K. (2000). A review of Jatropha curcas: an oil plant of unfulfilled promise. *Biomass and Bioenergy, 19*, 1–15.

Polatidis, H., & Haralambopoulos, D. (2004). Local renewable energy planning: A participatory multi-criteria approach. *Energy Sources, 26*(13), 1253–1264.

Pouliot, M. (2012). Contribution of "Women's Gold" to West African livelihoods: The case of shea (*Vitellaria paradoxa*) in Burkina Faso. *Economic Botany, 66*, 237–248.

Rahut, D. B., Ali, A., & Behera, B. (2015). Household participation and effects of community forest management on income and poverty levels: Empirical evidence from Bhutan. *Forest Policy and Economics, 61*, 20–29.

Raman, S., Mohr, A., Helliwell, R., Ribeiro, B., Shortall, O., Smith, R., & Millar, K. (2015a). Integrating social and value dimensions into sustainability assessment of lignocellulosic biofuels. *Biomass and Bioenergy, 82*, 49–62.

Raman, S., Mohr, A., Helliwell, R., Ribeiro, B., Shortall, O., Smith, R., & Millar, K. (2015b). Integrating social and value dimensions into sustainability assessment of lignocellulosic biofuels. *Biomass and Bioenergy, 82*, 49–62. doi:10.1016/j.biombioe.2015.04.022.

Rocheleau, D., Thomas-Slayter, B., & Wangari, E. (1996). Feminist political ecology: A feminist political ecology perspective. In D. Rocheleau, E. Wangari, & B. Thomas-Slayter (Eds.), *Feminist political ecology: Global issues and local experience* (pp. 3–23). New York: Routledge.

Rocheleau, D., Thomas-Slayter, B., & Wangari, E. (2006). Gender and environment: A feminist political ecology perspective. In N. Haenna & R. Wilk (Eds.), *The environment in anthropology: A reader in ecology, culture and sustainable living* (pp. 27–33). New York and London: New York University Press.

Romijn, H., Heijnen, S., Colthoff, J., de Jong, B., & van Eijck, J. (2014). Economic and social sustainability performance of Jatropha projects: Results from field surveys in Mozambique, Tanzania and Mali. *Sustainability, 6*(9), 6203–6235. doi:10.3390/su6096203.

Rosso-Cerón, A. M., & Kafarov, V. (2015). Barriers to social acceptance of renewable energy systems in Colombia. *Current Opinion in Chemical Engineering, 10*, 103–110.

Schoneveld, G. C., German, L. A., & Nukator, E. (2011). Land-based investments for rural development? A grounded analysis of the local impacts of biofuel feedstock plantations in Ghana. *Ecology and Society, 16*(4), 10.

Schoneveld, G. C., & German, L. (2014). Translating legal rights into tenure security: Lessons from the new commercial pressures on land in Ghana. *The Journal of Development Studies, 50*(2), 187–203.

Schoneveld, G. C. (2014). The geographic and sectoral patterns of large-scale farmland investments in sub Saharan Africa. *Food Policy, 48*, 34–50.

Schut, M., Cunha Soares, N., van den Gen, G., & Slingerland, M. (2014). Multi-actor governance of sustainable biofuels in developing countries. *Energy Policy, 65*, 631–43.

Selman, P. (2004). Community participation in the planning and management of cultural landscapes. *Journal of Environmental Planning and Management, 47*(3), 365–392.

Slingerland, M., & Schut, M. (2014). Jatropha developments in Mozambique: Analysis of structural conditions influencing niche-regime interactions. *Sustainability, 11*, 7541–63.

Suwelack, K., & Wüst, D. (2015). An approach to unify the appraisal framework for biomass conversion systems. *Biomass and Bioenergy, 83*, 354–365. doi:10.1016/j.biombioe.2015.10.012.

Timko, J. A., Amsalu, A., Acheampong, E., & Teferi, M. K. (2014). Local perceptions about the effects of Jatropha (*Jatropha curcas*) and Castor (*Ricinus communis*) Plantations on households in Ghana and Ethiopia. *Sustainability, 6*, 7224–7241.

van Eijck, J., Romijn, H., Smeets, E., Bailis, R., Rooijakkers, M., Hooijkaas, N., & Faaij, A. (2014). Comparative analysis of key socio-economic and environmental impacts of smallholder and plantation based jatropha biofuel production systems in Tanzania. *Biomass and Bioenergy, 61*, 25–45. doi:10.1016/j.biombioe.2013.10.005.

von Maltitz, G., Gasparatos, A., Fabricius, C, Chittock, A., & Willis, K. (2016). Jatropha cultivation in Malawi and Mozambique: Ecosystem services tradeoffs and their effect on local human wellbeing and poverty alleviation. Submitted to Ecology and Society.

von Maltitz, G. P., Gasparatos, A., & Fabricius, C. (2014). The rise, fall and potential resilience benefits of Jatropha in Southern Africa. *Sustainability, 6*, 3615–3643.

Walker, G. (2008). What are the barriers and incentives for community-owned means of energy production and use? *Energy Policy, 36*(12), 4401–4405.

Walker, G., Devine-Wright, P., Hunter, S., High, H., & Evans, B. (2010). Trust and community: Exploring the meanings, contexts and dynamics of community renewable energy. *Energy Policy, 38*(6), 2655–2663.

Wüstenhagen, R., Wolsink, M., & Bürer, M. J. (2007). Social acceptance of renewable energy innovation: An introduction to the concept. *Energy Policy, 35*(5), 2683–2691.

Index

A
Abandonment, 158
  of communal plantations, 190
  of community-owned properties, 172
  of private houses, 161
  of rice terraces, 187, 191
Academia, 42, 62
Academic field, 23
Academic recommendations, 132
Access, 161
  to basic services, 162
  to care services, 170
  to groceries, 170
  to PES, 44
  to transportation, 161
Accredited NGO, 53
Acetaldehyde, 95, 97, 98, 101, 107
Acquisitions, 200
  construction work permit, 204
  land, 200, 206, 208, 211, 222
  large-scale, 206, 212
  operational license, 204
  provisional license, 204
  siting clearance, 204
Activities, 78, 81, 83, 85, 86, 89
  community, 156, 172
  economic, 159
  local, 157, 166
  seasonal, 164
Adaptation strategies, 119
Additional part-time job, 167
Adjustment strategy, 178
Aeta, 182, 188–191
Afforestation of pasturelands, 50
Africa, 6, 200, 219
  Sub-Sahara Africa (SSA), 200–203, 222
  livelihoods, 201, 206, 212–214, 216, 219
Agano River, 98
Agenda of governments, 127
Aging
  society, 129
  at communal level, 172
  in Japan, 154
  in rural areas, 155
  population aging, 154
Agrarian
  community, 209, 215
  urbanism, 123, 126
Agriculture, 201, 209, 211, 212, 215, 219
  fallow land, 212, 214, 219, 222
  lands, 44, 46, 50
  plantation, 200, 211, 213–215, 219
  smallholder, 200, 201, 213, 219
  stakeholder, 201, 205, 220, 222
Agro-ecosystems, 45
Agroforestry
  development, 42, 44
  experts, 55
  farms, 42, 55, 56, 59, 61
  projects, 47
  systems, 44, 50–52, 55, 56, 58, 59, 62
Akita, 154, 158
Alternative
  concepts, 123
  livelihoods, 185
America, 118
Analysis, 32, 34, 35
  of secondary data, 119
  qualitative, 81
  quantitative, 80
Ancestral sovereignty and self-determination, 178

M. Esteban et al. (eds.), *Sustainability Science: Field Methods and Exercises*,
DOI 10.1007/978-3-319-32930-7

Anecdotal evidence, 127
Angle of observation, 147, 149
Anglo-saxon model, 122
Anticipatory competence, 70, 71, 77, 83, 84
Applications, 149
  process, 52
Appraisal approaches
  bottom-up sustainability, 220, 221
  rapid sustainability, 201, 205, 206, 208, 220–222
  reductionist, 205
  top-down sustainability, 220, 221
  unified sustainability, 220
Approach, 24, 25, 27, 28, 34, 37
Arid, 24, 27, 28, 30
Artifacts, 143
Asia, 6, 11
Assessment, 199
  bioenergy projects, 219, 220, 222
  Environmental Impact Assessment (EIA), 199, 202, 205, 206
  sustainability impact, 205, 220, 221
Assessment Discussion of Tohoku Unit 2015, 85
Assessment methodology
  Contribution to Competency, 79
  qualitative analysis, 80, 81
    students group discussion, 81
  quantitative analysis, 80, 81
    active and passive learning, 80
    survey design, 80
Assistance, during recovery process, 155
Ataxia, 95
Atmosphere and Ocean Research Institute (AORI), 76, 83–85
Australia

B

Backcasting, 178
Background, 142, 144, 146
  layers, 145, 147
Baseline assessment, 62
Basic services, 153, 161
  access to, 162
Bays, 72
  Funakoshi Bay, 72
  Otsuchi Bay, 72
'Before' and 'after' maps, 183
Beijing, 129
Benefit sharing, 205, 208, 212
BioFuel Africa, 214. *See also* Biofuels
Biodiversity, 42, 44, 46, 49, 50, 55, 60
  loss, 3
  protection, 42
Bioenergy project/system, 205, 206, 220, 222
Biofuels, 200, 201, 205
  company/investor, 203–205
    Galton Agro Ltd, 207, 210, 217
    Kimminic Estates Ltd, 207, 210, 217
  policy, 200, 203, 220
  projects, 202, 206, 211, 216, 220, 222
    community participation, 202, 203
Biology, 7
Biomagnification, 99, 100, 108
Biomass
  biofuels, 221
Bird-eye view, 141
Blog posting, 93, 103, 106, 111
Blueprint, 42
Bottleneck, 44
Bottom-up approach, 7, 11
Boundary, 32–34
Brundtland, 129
Building
  building types, 126
  building block, 125, 126
  building plots, 125

C

Capability
  of individual communities, 153, 156
Career, 161
Cash remuneration, 42
Cat #400, 97
Categories
  of survey questions, 160, 161, 170
Centreless city, 125
Certificate of Ancestral Domain Claims (CADC), 180, 182
Certificate of Ancestral Domain Title (CADT), 180–182, 189, 191
Certificate of Ancestral Lands Claims (CALC), 180
Certificate of Ancestral Land Title (CALT), 180, 181
Certification process, 101, 109, 219
Challenges
  in property management, 166
  local knowledge management, 173
  of rural communities, 158
  property management, 172
Changes
  in rural communities over time, 156
  of target system, 156

within rural communities, 154
Chiefs, 200, 205, 212, 215
Chieftaincy trust, 215
China, 26, 28, 30
environment problems
Chisso, 95, 97, 98, 102, 108, 111
Christianization, 186
City
analysis, 122
beautification, 118, 131
functions, 118
Civil societies, 127
Clear evidence
of community marginalization, 171
of population aging, 154
Climate change, 3, 6, 13, 118, 120, 123, 130, 132
adaptation to, 119
Clinical issues of Minamata disease, 109
Co-design, 125, 130
Coffee farmers' cooperative, 48
Collaboration, 4, 6, 121
with local partners, 45
Collaboration Center for Minamata Disease Victims in Minamata, 104
Collapsed jatropha projects, 201, 205, 206, 211, 212, 215, 221
feedstock selection, 200, 203
financial planning, 200
investor funding, 200
mismanagement, 200
mitigation strategies, 200, 205, 219, 221
Collective, 25, 26, 33
management, 178, 192
Combination of aging and depopulation, 154, 158
Commercial activities, 125
Common challenge
in property management, 166
Common type of farming, 165
Communal ceremonies, 157
Communal irrigation system maintenance, 191
Communal land, 187–189
Communal level, population aging at, 172
Communal plantations, 190
Communication, 27, 32
Community, 72, 76, 77, 201, 203, 204, 213, 215, 216, 219, 221
acceptance, 201–203, 208
networks, 87
participation, 201
conceptualizing biofuel project, 202, 203
distributive justice, 208
procedural justice, 206, 208, 222, 223
redistributive justice, 203, 206, 222, 223
perceptions, 205, 206, 208, 211, 219, 220, 222
social service, 201, 214
Community-Based Forest Management (CBFM), 180–183, 189
changes through, 186
implementation, 181, 182, 189
in Lingay, 185
mediocre outcomes of, 191
role, 190
Community empowerment, 180
Community-function, 155, 157
Community groups, 160
Community marginalization process
community groups in, 171
in Kamikoani, 170
in rural areas, 155, 156, 158
stages, 157
Community vitality, 156
Community workshop, 188
Compact city, 126
Compensation, 100, 102, 106, 108, 110, 111, 205, 208, 211, 212, 222
Competence, 34, 35, 81
anticipatory, 83
interpersonal, 84
normative, 84
strategic, 84
system-thinking, 83
Complex design, 121
Complex dynamics, 139, 144, 146
Complex issues, 3, 7
Complexity, 5, 7, 9, 11, 15, 25, 28, 37, 138–141, 143, 147, 149
Complexity thinking, 139, 140
Complex systems, 140
Comprehensive concepts, 123, 126
Concept, 142
of community marginalization, 157
of resilience, 119
Conceptual model, 147
Conceptual understanding, 137
Conditionality, 44
Conditions
in marginal communities, 158
of community, 157
of households, 170
of residents, 172
Congrès Internationaux d'Architecture Moderne (CIAM), 121
Connections, 146, 149

Consensus building game, 111
Conservation, 142, 146
Consumption. *See* Self consumption
Context, 138, 143, 145, 147–150
Contextual situation, 117
Contextual understanding, 139, 144
Contextualization, 139
Continuous depopulation, 158, 171
Contracts, 49, 50, 53
Coope Dota, 58
Costa Rica, 15, 42, 43, 45–48, 56, 58, 60–62
Cost-efficient ways for ecosystem service provision, 62
Course, 24, 31
Creations, 143
Creative output, 102
Criteria selection, 52, 199
Cultural, 25, 33–35, 37
Cultural landscape, 179, 182
Culture, 122, 161
Curriculum, 5, 6, 24, 25
Customary law, 182

D
Daily products, 164
Dalligan, 182, 184–186
Damage, 72, 73, 75, 76
Death and life of great american cities, The, 123, 125
Decentralization, 202, 203. *See also* Project planning
Decision makers, 124, 127
Decision model, 138, 149
Decisions, 119, 138
Decline
  economic decline, 154
  forestry decline, 160
  of community-functions, 171
  population decline, 153, 156, 157
Deforestation, 42, 46
Degradation, 29, 30
Degree
  of community-functions, 157, 171
Demographic profiles, 119
Demographic trends, 153, 154
Department of Environment and Natural Resources (DENR), 180, 182
Depopulation, 154–156, 158, 171
De portzamparc, Christian, 121
Desertification, 29
Design, 24, 27, 30, 34, 35, 37
  activities, 83
  concept, 131
  juries, 117
  of field exercise, 81
  of field units, 89
  of survey, 80
  phase, 124
  profession, 124, 126, 132
De solà-morales, Manuel, 121
Determination of Capacity use of Lands in Costa Rica, 52
Development, 205, 214, 215, 220–222
  countries, 119
  of competencies for sustainability research, 77
  of anticipatory competence, 83
  of competencies, 89
  of interpersonal competency, 85
  of strategic competency, 84
  paths, 191, 192, 194
  rural, 200–202, 204, 222
  urbanl, 202
Dialogue, 147, 148, 150
Differentiation, 140, 147
Dimension, 35, 37, 38
  economic, 123, 127
  environmental, 123, 127
  of target system, 154
  of sustainability, 85, 89
  social, 123, 127
Disaster, 73, 75, 76, 86, 87, 89, 130, 132
Disciplines, 121, 122
Discourse, 137, 139, 161
  on leadership
Discussion, 28, 37
Dissatisfied participants, 186
Dissemination of information, 112
Domain
  science and engineering, 128
Drinking water supply, 187, 192
Dynamic patterns, 139, 144, 149, 150
Dynamics, 139, 144

E
Earthquake, 73, 75, 76, 131
Eco-city, 123, 129
Ecology, 205, 206, 220. *See also* Environmenatal impacts
  issues, 68, 83
  urbanism, 123, 124, 129
Ecomarkets I, 49
Economic
  concept in urban planning, 125

commodities, 44
development, 144
disadvantaged areas, 50
incentives, 47
income satisfaction level of households, 167
issues, 81
state, 161, 162
status, 170
Ecosystem, 29, 32
based adaptation, 123
services, 124
Education, 4, 6, 15, 24, 25, 27, 34, 37, 187
game development, 112
programs, 69
purposes, 70
sustainability program, 70
Educational and research field methodologies, 75
applications, 75
assessment results, 81
development, 77
study area, 72
sustainability in reconstruction process, 77
Empirical knowledge, 45
Employment, 209, 211, 213, 216, 219
End-state, 130
Energy, 4, 220. *See also* Bioenergy project/system
energy security, 200, 222
renewable energy, 201, 202, 204
Engineer, 32
Environment, 26, 27, 29–33, 155, 162, 168
benefits, 44
consciousness, 44
degradation, 49
development, 15, 16
education, 15
impacts, 208, 213, 216, 219, 222
air quality, 217, 222
soil quality, 213, 216
water quality, 213, 216
issues, 11, 68, 81, 85
leadership, 15
problem, 11
service sellers, 62
services, 42, 49, 50, 52, 55, 60–62
law, 48
Europe, 118, 122, 129
Evaluation, 138, 139, 142, 143, 146, 147, 149, 150, 205, 221
on current income levels, 167
Events
local events, 156, 157, 166, 171
Evidence
of community marginalization, 170
of population aging, 154
Evidence-based, 126
Evidence-based design, 123
Evidence-based problem-solving, 131
Executive decree, 52, 61
Executive Decree No. 23214 - MAGMIRENEM, 52
Exercise on Resilience (ER), 15, 68, 69, 72, 75, 80, 86, 87
Experience, 28, 30, 35, 38, 77, 156, 158
Expert and local knowledge, 147
Exploratory, 42
Exterior, 25, 26, 33
Extinction, 61

F

FONAFIFO regional officer, 53
Facilitators and implementer, 42, 46
Fact
continuous depopulation, 171
Faculty, 30, 32, 75, 89
Failures of community forestry, 190
Family, options for, 168
Farming, 123, 164, 165, 170
Farmlands, 158, 164
Farm to market road, 187
Feedback, 7, 144, 145
Feedstock, 200, 203, 205, 206, 219, 222
biofuel production, 200, 203, 205, 212, 215, 219, 220, 222
poor selection, 200
Fertility rate, 154, 155
Festivals, 157, 166
Field-based method, 193
Field exercise, 5–7, 9, 11, 15, 24, 25, 27, 28, 31, 34, 37, 138, 150
Field level survey, 42, 46
Field method, 138, 144, 147, 150
Field research, 117, 127, 144, 147, 150
Fields of urban planning
disciplines, 119
practitioners, 119
Field study, 42
Field survey, 45, 47
Fieldwork, 7, 15, 24, 30, 32, 37, 47
Financial insecurity, 4
First Forest Law, 48
Fishermen, 77, 83
Fishermen's catch, 100, 108

Fishery industry, 77, 83
Fishing, 76, 77
Five elements
  districts, 122
  edges, 122
  landmarks, 122
  nodes, 122
  paths, 122
Flemish Zoning Code, 129
Flexible law, 61
Focus, 141, 142
Focus–system, 142, 146
Food
  availability, 212, 214, 216
  production, 219
  security, 201, 206
Foreign direct investment (FDI), 200, 206
Forest, 159, 164, 172, 206, 213, 214, 219
  cover, 42, 47, 54
  decentralization, 192
  non-traditional forest products (NTFPs), 213, 215, 216
  protection, 42, 47, 49, 50
  recovery, 48
  restoration, 47
Forest cover target (2020), 55
Forestry, 159
  incentives, 42
  plan, 52, 56, 59, 62
Formal land title, 52
Formalist, 118
Formation
  of local groups, 153
Foundational reconstruction, 73
Four-quadrant, 24–26, 33
Framework, 69, 80, 87
  marginalization framework, 154, 157, 158, 160, 171
Frequency of grocery shopping, 164, 170
Full-fledge reconstruction, 75
Functions
  community-functions, 153, 156, 157
  of local communities, 155
Future demography, 153
Future planning, 147, 150
Future sustainability, 71, 83

G

GIS tracking, 47
GPSS-GLI International Symposium, 46
GPSS-GLI. *See* Graduate Program in Sustainability Science– Global Leadership Initiative (GPSS-GLI)
Garden cities of to-morrow, 123
Garden city, 124
Gatherings, 156, 157
Gender differentiation, 208, 215
Generations, 173
Gentrification, 131
Geographers, 122, 130
Geographical spaces, 179
Ghana, 205, 222
  Adidome, 209, 212, 213, 215
  biofuel project, 204–206, 219
  Kobre, 209, 212, 213, 219
  Kpachaa, 209, 212–214, 219
  government, 200, 203, 206
  income, 211, 214
  literacy rate, 213, 215
  local communities, 201
Global environmental problems
Global Field Exercise (GFE), 5, 6, 13, 15, 42, 43, 45
Global Leadership Exercise (GLE), 5
Global, 24, 25, 30, 32
Globalization, 178, 192, 194
Global population, 154
Glocalization, 178, 192, 194
Good city form, 122
Gosho-no-ura Island, 104
Governance, 143
Governing practices, 143
Graduate Program in Sustainability Science–Global Leadership Initiative (GPSS-GLI), 68, 69, 72, 75, 90
Green ecosystem layer, 123
Groceries, 162
  access to, 162, 170
  mobile grocery stores, 164
Group discussion, 77, 79–81, 83, 84, 88
Groups
  community groups, 153, 159, 160, 162, 168, 170, 171
Group work, 6, 15, 85
Guanacaste Conservation Area (ACG), 47
Guanacaste province, 47

H

Hachiman pool, 97
Hard infrastructure, 123, 129
Hard solution, 125
Heihe River basin, The, 15
Historical analysis, 146
Holistic, 4, 6, 9, 11–13, 24, 25, 30, 33, 37
Holistic thinking, 138, 139, 142, 144, 145, 149, 150

Households, 153, 160, 162, 164, 169
conditions of, 170
income satisfaction level of, 167
livelihoods of, 172
operation of, 165
part-time farming, 170
private properties of, 172
property management of, 164
proportion of, 170
survey, 202, 206, 208, 216, 221
Houses, 158, 161, 164
Howard, Ebenezer, 124
Human geographers, 119
Humanity, 3
Human–natural systems, 138–140
Human scale, 124
Hunter-Russel syndrome, 95
Hyakken port, 97

I
Ifugao, 182, 183, 186
Image of the city, 122
Impact, 27, 29, 33
of population aging, 154, 172
of social dimensions, 154
Impaired hearing, 95
Implementation structure, 42, 43
Incentives for reforestation, 48
Inclusive design, 60
Income, 153, 167, 168, 170
Income groups, 126
Income tax reductions, 48
Increase the accessibility of PES to smallholders, 42
In-depth investigation, 45
In-depth literature survey, 42
Indicator, 220
Indigenous
cultural communities, 142, 177, 178, 180, 182, 191
law, 178
people, 178, 180, 183, 194
population, 52
Individual, 25, 27, 33
community, 156, 170
from rural to urban areas, 155
households, 164
Industrial pollution, 101, 106, 108
Infill development, 129
Informal land ownership, 44
Informal settlements, 122
Inhabitants, 117, 146, 182, 187, 188, 208
Innovation, 143
Institution, 27, 30, 37
Intangible resources, 164
Integral approach, 15
Integral framework, 28, 33, 37
Integrated, 56
Integrated Research System for Sustainability Science (IR3S), 4
Interaction, 144
among residents, 170
with outside communities, 172
with urban residents, 158
Integration of trees, crops and animals, 44
Interdependency
Interdisciplinary, 24, 25, 37, 121, 127
research, 11, 148, 150
Intermediaries, 51, 53
Intermediary
agencies, 53, 56, 60
Interior, 25, 26, 33
Intermingling of housing, 126
International Union for Conservation of Nature (IUCN), 48, 60, 62
International community, 130
International organizations, 119
Interpersonal competence, 70, 71, 81, 84
Interpretation, 139, 140, 143, 147
Interregional
Interrelationships, 121
Interviews, 33, 72, 76, 79, 83
group sizes, 86
semi-structured, 85
Irrigation, 182, 187, 191, 192
Issue, 36, 37, 142, 144, 146
in rural planning, 161
of local economy, 162
related to residents' mobility, 170
rural issues, 158
Iteration, 143
process, 45
understanding, 143, 150
Iwate, 72, 73, 76, 83, 84

J
Jacobs, Jane, 124, 129
Japan, 7, 13, 15, 16, 30
population aging in, 154
rural studies of, 156, 161
Japanese Bank for International Cooperation (JBIC), 182
Japan International Cooperation Agency (JICA), 182
Jatropha, 200, 201, 203, 212, 213, 219. *See also* collapsed jatropha project
Jobs, 167
and incomes, 212, 213

K
Kajikawa, 128
Kamikoani village, 154, 158–160, 165, 171, 172
Key competency, 69, 70, 80, 87, 89
Key stakeholders, 42
Kiangan, 182, 184, 190, 191, 194
Knowledge, 4, 10–12, 15, 24, 25, 27, 30, 34, 68, 69, 81, 87, 164, 173
  economic, 126
  local, 173
  scientific, 126
  social, 126
Knuckels forest reserve, 146
Kumamoto Prefectural Government, 104, 105
Kumamoto Prefectural Government Division of Minamata Disease, 104
Kumamoto Prefecture, 94–96

L
La Gaceta No. 143, 52
Labor shortages, 154
Lack of additionality, 45
Land, 76, 84
Land cover change, 49
Landowners, 42, 50–52, 55, 56, 59–62
Land readjustment (LR), 123, 126
Large-scale agriculture, 119
Landscape
  architects, 124, 126
  ecologist, 124
  urbanism, 123, 124
Land use changes, 49
Land use modalities, 42, 50, 55
Leadership, 4, 5, 11, 15, 24, 25
  experience, 5
  development, 4, 6, 11, 15
  education, 15
Learning, 70, 80
  active, 80
  passive, 80
  traditional modes, 89
Le Corbusier, 121
Lectures, 72, 75, 77, 81, 83–86, 89
Legal easement, 52
Legal proof of land ownership, 44
Levels
  communal level, 155, 172
  community level, 160
  income satisfaction levels of, 167
Levels of expected learning, 75, 77
  competencies development, 70
  outcome of field work, 70
Levittown, 123, 125
Licensed forest engineer, 53
Limited prior knowledge, 42
Lingay, 182, 184–186, 194
Literature reviews, 42, 45, 46
Liveability, 131
Live hedges, 50
Livelihoods, 161, 162
  in rural communities, 168
  of households, 172
Living environment, 155, 158, 168, 170–173
Local activities, 157, 166
Local culture, 161
Local events, 166, 171
Local festivals, 166
Local forest resources, 178, 182
Local government, 73, 118, 129
Local groups, 153
Localization, 147
Local knowledge, 164
Locally relevant issues, 45
Long-term challenge, 118
Long-term mechanism, 123
Long-term sustainability, 61
Low-cost housing, 125
Low-intensity logging, 50
Luxury building, 124
Lynch, Kevin, 122

M
Machizukuri, 123, 125, 129
Maintenance
  for individual households, 165
  of community-functions, 171
  of community properties, 161
  of living conditions, 171
  of local living, 155
Major tropical systems, 47
Malnourishment, 190, 191
Management
  of households, 164
  of local resources, 156, 173
  of private properties, 173
  property, 161, 166, 172
Manpower
  for local activities, 166
  shortage of, 167
Marginalization framework, 154, 156, 158, 160
Market, 158
Market-oriented approach, 50
Mass-produced urbanisation, 125
Matrix method, 148
Measurability, 126

Meemure, 146, 149
Members, 157, 166, 168
Mercury-contaminated fish, 99, 100, 108
Mercury pollution, 95, 99
Method, 24, 27, 31
Methodologies, 24, 26, 27, 31–33, 37, 42, 43, 45, 52, 120, 122, 128
  approach, 45
  challenge, 154
  in Exercise on Resilience (ER),Education and Research, 75, 77
    guide tour, 76
    interviews, 76
    lectures, 75
    social gathering, 77
  of field exercise contribution to competencies, 79
  for rural areas sustainability examining, 156
  in community marginalization, 156
Methyl mercury, 95, 97–99
Migration
  to rural areas, 162
  to urban areas, 154
Millennium Ecosystem Assessment, 45
Mimaikin Solatium Agreements, 100
Mimaikin (sympathy money), 100
Minamata Bay, 96, 97, 100, 108
Minamata City, 94–96, 98, 101, 102, 104, 107
Minamata Disease, 7, 15, 95–98, 101, 108, 111
Minamata Disease Victims Mutual Aid Association, 104
Minamata Disease Patients Alliance, 104
Minimum land area, 44
Ministry of Environment and Energy (MINAE), 48, 52, 53, 61
Mitigation of greenhouse gas emissions, 42
Mixed methods, 131
Model, 25, 33
Mode-1 science, 126
Mode-2 science, 126
Modernistic planning theory
  lacks flexibility, 119
Monitoring and compliance, 53
Modernism, 118, 121
Monodisciplinary sustainability science, 126
Morphological
  layout, 125
  methods, 129
  planning, 125, 127
  shapes, 128
Mozambique, 200, 201
Mt. Pinatubo, 188
Multi-functional agroforestry system, 60
Multiple stakeholders, 43
Multiple views. *See* observation methods

N
National Biodiversity Institute (inbio), 48
National Census surveys, 155
National Forest Financing Fund (FONAFIFO), 49, 50, 52, 53, 55, 56, 60, 62
National level
  population aging at, 154
National parks, 48
National System of Conservation Areas (SINAC), 47, 53
Native and foreign tree species, 61
Natural, 25, 26, 31, 33
  disaster, 13
  interaction, 128
  typography, 124
Nature, 138, 146
  degradation, 142, 146
  nature based solutions, 123
Negative environmental externalities, 49
Neighborhood-based social ties, 155
New research project, 45
New urbanism, 123, 126
Niigata Prefecture, 98
Nippon Nitrogen Fertilizer Corporation, 95, 104
Nitric acid, 95, 97, 98
Normative competence, 70, 71, 84
Non-governmental organization (NGO), 25, 48, 51, 53, 60, 62, 100
North America
Number of trees planted, 42, 55, 60, 61

O
Oasis, 24, 26–30, 32, 35–37
Observation methods, 138, 139, 141, 143
  dimensional view-based method, 147
  layer view-based method, 146
Oil palm, 200, 206
One third of the sales tax on fossil fuel, 49
On-site, 24, 34
Open space, 124
Operation of households, 165
Opportunistic behaviour by landowners, 62
Optimal provision of environmental services, 42

Optimization, 55
Organic growth, 124
Organizational role, 48
Organizing relationships, 140, 144, 145
Otsuchi Fukko Suishin Tai (NPO), 78, 83, 86
Otsuchi Town, 70, 72, 73, 75, 81, 84, 85, 89
Outcome, 27, 34, 35, 37
Outside of communities, 171–173
Overarching understanding, 139
Overpopulation, 121

P

PES
  for agroforestry, 42, 44, 46, 53, 54, 60, 61
  modalities, 42, 44, 50, 51
Part-and-whole, 140
Participation
  appraisal, 180, 192
  approach, 125
  development, 177, 179
  mapping, 177, 179, 180, 184, 190, 192
Parts, 140, 141, 148
Passive learning, 80, 83, 85, 87
Patient and sufferer, 109, 111
Payment for ecosystem services, 44
Payment scheme, 44, 51, 55, 60, 61
Payments by hectare, 42
Payments for Environmental Services (PES), 42–46, 48–50, 52, 54, 56, 58, 60–62
Pedestrian permeability, 125
People, 29, 30, 33, 38
  migration, 155
Peoples' Right Act, 180
Performance-based, 60, 61
Period, 153, 155, 156
Permanganate, 95, 97, 98
Person, 165
Personal competence, 87
Perspectives, 154, 155, 158, 171, 173
Perverse incentive, 55, 60
Phenomenon, 155, 158
Philippines, 180
  case study in, 181
  in CBFM projects, 194
Philosophy, 5, 7, 12, 13, 140, 150
Physical
  changes, 125
  shape, 125
  solutions, 118
Pillars of sustainability, 9
Place, 155, 156, 170
Planet, 139
Planetary boundaries, 32–34
Planetary wellbeing, 142
Planning, 161
  acquisitions, 122
  compact city, 118
  design, 118
  physical space, 118
  policy, 129
  systems, 121
Planting
  density, 55, 56
  distance, 56
Policy, 30, 32–34, 143, 147
  focus on identifying gap, 120
  makers, 126
  objectives, 42, 43
  policy contexts, 118
  results, 43
  variety of policy, 118
Population, 73, 77, 83, 85
  aging, 7, 154
  change, 4
  declines, 153
  impacts of, 172
  in Japan, 154
  in Kamikoani, 158
  in rural areas, 155
Position
  of community, 158
Positive externalities, 50
Post-colonial movements, 179
Post-fieldwork, 34, 35
Poverty, 4, 144, 146
  reduction, 50
Practical, 24, 30, 31
Practice, 24, 25, 29, 37, 144, 146
Practitioners, 42, 126, 138, 144, 148
Prefabrication, 125
Prefecture
  Akita, 154, 158
Preliminary assessment, 42
Preliminary surveys, 6
Prioritization criteria, 61
Priority, 173
Pritzker prize, 121
Private
  landholders, 48
  properties, 161, 172
  woodlots, 187
Problem-based, 45
Problem-oriented perspective, 5
Problem Ranking, 177, 180, 182, 184, 192, 193
Problems, 156, 168
Problem solving, 43, 127

Process. *See* Community marginalization process
reconstruction process, 69, 70, 73, 76, 77, 83, 84, 86, 87, 89
recovery process, 68, 72
Processing facilities, 77, 83, 84
Production, 157
Profession, 121
Program, 24, 27, 30, 34, 37
Programme, 69
Project, 25, 30, 32
implementation, 200, 201, 203, 204, 208, 214, 219
CBFM, 189, 193
design, 201, 203, 216, 219–222
project planning, 201–203, 213, 220, 222. *See also* Biofuels
decentralization system, 202
stages, 204, 206
Proof of ownership, 44
Properties
community, 161
community-owned, 172
in rural areas, 164, 172
local, 158
private, 161
Proportion, 154, 155, 158, 165, 167
of households, 170, 173
Proposed framework, 158
Protected forests, 49
Protection of water, 42
Provision of scenic beauty, 42
Proximity, 129
Public facilities, 170
Public realm, 128

Q
Quadrant, 25, 26, 28, 31, 34, 38
Quadrivia, 25, 27, 32
Qualitative changes in rural communities, 153, 156
Qualitative methods, 47
Qualitative survey, 117
Quantitative survey, 117
Quality
of agroforestry farms, 42, 55
of community-function, 156–158
of life, 10
of trees, 60
within rural communities, 154
Questionnaire, 80, 81, 84, 87, 153, 160, 162, 164, 168
Quota for the number of hectares and trees, 52

R
Rapid appraisal. *See* Appraisal approaches
Rapid assessment, 192, 194
Rate
fertility rate, 154, 155
Recommendations, 130
Reconstruction issues, 75
Reflexive understanding, 150
Reflexivity, 143, 150
Reforestation, 42, 48, 50, 182, 185, 186, 190, 191
subsidies, 48
Region, 24, 27, 28, 32, 38
Tohoku region, 75
Regional activities, 33, 161
Regrowth of tropical dry forest, 47
Regulations, 143
Relationships
among residents, 168
social, 161, 162
Relief of Pollution-Related Health Damage, 101
Religious practices, 186
Remaining community members, 157
Removing snow, 165, 166
Research, 156
researcher, 25, 27, 37, 38
research field methodologies, 75, 85, 87
research questions, 45, 46
Residential housing project, 129
Residents, 75, 77, 85, 155, 157, 158, 162, 168, 170–173
Resilience, 131, 142, 192
aspects, 68, 87
thinking, 14
resiliency, 13, 14, 16
resilient society, 4, 6, 7, 13, 14
Resistance, 178
Resources, 27–29
depletion, 118, 128, 130
intangible, 164
limitations, 143, 144
local, 156, 173
Respondents, 162, 164
Responses, 166–168, 170
Review, 24, 32, 33
Revised Forestry Law 7575, 49
Revitalization, 162
local economy, 158
local festivals, 170
Rice terraces, 182, 187, 192
Road connection, 187
Rules, 143
Rural areas

Community marginalization, 155
sustainability examination methods, 156
Rural communities
sustainability of, 172
Rural issues, 158

S
Safety
of residents, 123
Salmon fishing, 73, 77, 83
San Jose city, 47, 55
Satisfaction
in income, 168, 171
level of households, 167
Scenarios, 141, 150
Science direct search engine
sustainability, 120
urban design, 120
urban planning, 120
Science of Minamata Disease, 106
Science practitioners, 130
Science. *See* Sustainability science
Scientific knowledge, 131
Scientific papers, 120
Scientific reality, 117
Scientific research, 127, 130
Seasonal gatherings, 157
Second-generation, 42
Selection process, 52
Self-consumption, 165, 170, 173
Self-governed areas, 179
Self-organization, 140
Semi-Nomadic Tribe, 182, 188
Sensory disorder, 95
Service buyer, 44
Service provider, 44
Settlement, 95, 101, 106, 110, 111
Shade coffee, 50, 58
Shares
of households, 168
of part-time farmer, 165
of self-consumption farming type, 165
Shifting cultivation, 188, 189
Shiranui sea, 100, 102
Shopping, 164, 170, 171
Shortage
labor, 154, 165, 171
Showa Denko, 98
Single element of functionality, 125
Skill, 30, 31, 68–70, 85, 89
Slash-and-burn, 185, 186, 190, 192
Slow lane, 129
Small farmers, 158
Smallholders, 44, 51, 53
Small scale farming, 165
Snow, removing, 165, 166
Social
capital discourse, 162
changes, 186
dimensions, 154
discrimination, 102, 109
gathering, 77
inclusion, 125
issues, 81
observations, 125
phenomenon, 158
relationships, 161, 162, 168
among residents, 168
security systems, 154
wellbeing, 118
Societal transition, 43
Socio-cultural spaces, 179, 180
Socio-ecological systems, 142, 149
Socioeconomic challenge, 60
Socioeconomic impacts, 208, 209, 213, 214, 216, 222
Soft solutions, 125
Soft systems method, 147
Soil conservation, 44
Solid waste management, 187
Solution oriented framework, 45
Solution-oriented problem, 132
Sorensen, 125
Soshisha, 104
South America
Space, 143, 144, 146, 150
Spatial
analysis, 131, 132
characteristics, 122
issues, 131
Speech impairment, 95
Spread of diseases, 121
Sri Lanka, 146
Stages, in community marginalization, 157, 158, 171
Stakeholder, 6, 8, 11, 27, 30, 32, 75, 79, 83–85, 94, 105, 109–111, 128, 147, 150
engagement, 43
Start-up costs, 44
State
economic state, 162
economic states, 161
of community groups, 171
of rural community, 156, 157, 171, 173
State-driven scheme, 60

Status, economic, 167, 170
Stores
  mobile grocery stores, 162, 164, 170
Storm surge, 119
Strategic competence, 70, 71, 81, 84
Strauss, Lévi, 120
Structural Adjustment Program (SAP), 48
Structural separation, 58
Student, 24, 30–32, 37, 68, 72, 75–77, 79–81, 83–89
Study, 24, 26, 29, 32
Subjective, 26, 27, 33
Subsidized credits, 48
Subsidy for timber, 48
Subsystems, 146
Sugarcane, 200, 206
Support, external, 166, 171–173
Survey, 6, 11, 15, 24, 33, 81, 84, 88
  questions, 161, 162, 165, 166, 168
Sustainability
  achievement of a higher level, 124
  agriculture, 47, 51
  agroforestry, 56
  assessment, 139
  boundaries, 142, 148
  challenges, 118, 127
  community, 118, 119, 121
  community development, 179, 180, 183, 190–193
  contexts, 138, 142, 145, 150
  development. *See* economic development
  dimensions, 142, 144
  dynamics, 140, 146, 150
  forest management, 42, 50
  framework, 142, 149, 150
  indicators, 147
  issues, 70, 80, 87
  logging plan, 50
  methodology, 137, 142, 147
  of biofuel, 201, 219, 221. *See also* Appraisal approaches
    impacts, 201, 202, 205, 208, 220, 221
    science, 202, 205, 206
  of rural areas, 158
  of rural areas, examining, 156
  of social security systems, 154
  of target system, 156
  patterns, 139, 144
  policy, 147
  principles, 147
  problem, 43
  provision, 61
  science, 42, 43, 45, 69, 70, 77, 87, 88, 137, 144, 150
    planning concept, 120
    practical urban design, 120
  scientist, 42, 149, 150
  supply, 49
  sustainability-linked knowledge, 143
  sustainability-linked worldviews view changes, 143
  urban development, 16
Sustainable Development Goals (SDG), 119
Synergy, 58
Synthesis, 147
System, 138, 139
  environment, 140
  relationships, 142, 145, 149
  social security systems, 154
  systemic perspective, 158
  target system, 153, 156
  thinking, 146, 147
  view, 139, 140
Systems-thinking competence, 71

T

Tanzania, 200, 201
Target, 188–192, 194
  system, 153, 154, 156
  village, 182, 188
Task
  removing snow task, 165
Technical assistance, 53, 56, 58, 59, 62
Temporal, 7, 11, 14, 205
Temporary housing unit, 76, 77, 83, 85, 86
Thailand
Theoretical framework of Costa Rican PES, 43
Theoretical understanding. *See* Conceptual understanding
Theory, 138
Thresholds, 158
Ties
  social ties, 155
Tohoku Earthquake Tsunami (2011), 7, 15
Tohoku Unit, 68, 70, 75, 77, 87, 89, 131
Tokyo. *See* University of Tokyo
Top-down approach, 7, 9, 11
Totality of self-managing capability, 156
Tour
  bus tour, 84
  guided, 76, 86
Tourism, 47, 78, 168
Trade-offs, 214, 219, 221, 222
Traditions, 164, 173
  events, 156, 157
  farming, 45
  methodologies, 122

Trajectories of target system, 156
Transaction costs, 44, 53
Transdisciplinarity, 128
Transdisciplinary, 24, 26, 33, 37
research, 11, 147, 148, 150
Transferable bonds, 48
Transformation, 143
Transition, 43, 68, 126, 127, 158
Transportation, 161, 162, 171
of households, 163
Tree combinations, 55
Tree cover, 42
Trends
demographic trends, 153, 154
in population aging, 154
of agriculture and forestry decline, 160
of population decline, 155
out-migration trend, 155
Tropical Agricultural Research and Higher Education Center (CATIE), 48, 55, 56, 58–62
Tsunami, 7, 15, 68, 73, 75–77, 131
Tuwali, 182, 183
Types
of activities, 166
of challenges, 158, 172
of egg cooking
boiled egg, 122
industry, 122
modern city, 122
of properties in rural areas, 163

U
USA, 118
Uncertified patients, 111
Un-habitat, 118, 121
University, 28, 30, 32, 38
University of Lüneburg, 79
University of Tokyo, 68, 76, 78, 84
Urban
agriculture, 126
architecture, 121
areas
migration to, 154, 155
economic, 119
infrastructure, 125
landscapes, 121
morphological, 130
morphologists, 120
morphology, 130
paradigm, 119, 121, 124
problems, 16
social, 119
spaces, 122
sprawl, 125
Urban design, 118, 119, 121, 122, 124–126, 128–130
concept, 120
focused on morphology, 120
layers, 123
Urban environment, 16, 124
management, 16
problems, 16
Urbanization, 124, 128–130
Urban planning, 118–120, 122, 124–126, 128–131
field techniques, 119
foundation of contemporary, 121
layer, 123
morphological, 119
morphological analysis, 121
morphological research, 120
physical solutions, 119
planner, 119, 121, 122, 128, 130–132
problem-solving, 119
promote better environment, 121
solution-oriented, 124
typological, 119

V
Video production, 93, 109
Villagers, 170
Village. *See* Kamikoani village
Vision, 32
Visual field constriction, 95
Vitality
community vitality, 156
Vlaamse codex voor ruimtelijke ordening, 129
Volcanic eruption, 182, 188, 189, 192
Voluntary transaction, 44

W
Waldheim, 123
Waste water, 95, 97, 99
Water, 26, 28, 29, 31, 34, 35
supply, 187
tariff, 49
Well-being, 142
Well-defined environmental service, 44
Whole, 140, 141
Wholistic understanding, 141
Women headed households, 52
Working landscape, 44

World, 25, 29
World Bank, 48, 49
World demographic trends, 154

Y
Yokohari, 118, 130
Yonmenkaigi, 123, 125, 129

Z
Zhangye, 29, 32, 33
Zones, 118
Zoning, 118

FSC
www.fsc.org
MIX
Papier aus verantwortungsvollen Quellen
Paper from responsible sources
FSC® C105338

If you have any concerns about our products,
you can contact us on
ProductSafety@springernature.com

In case Publisher is established outside the EU,
the EU authorized representative is:
Springer Nature Customer Service Center GmbH
Europaplatz 3, 69115 Heidelberg, Germany

Printed by Libri Plureos GmbH
in Hamburg, Germany